Orbits and Climate Oscillations

by Rolf A. F. Witzsche

Contents

About the Illustrated Science series
On the Ice Age and Climate Change
and the book

Orbits and Climate Oscillations

Book 3 of the series: Ice Age of the Dimmer Sun in 30 Years

The orbit of the Earth around the Sun, and the orientation of the Earth's spin axis, oscillates slightly in intervals of 26,000, 41,000, and 100,000 years. These cycles, termed the Milankovitch Cycles, were once deemed to be the cause for the Ice Age cycles. This perception no longer holds true.

Evidence exists that the orbit of the Earth and the planets is actively maintained by the electrodynamic effects of the 'Primer Fields' that focus interstellar plasma onto our Sun, in times when they are active. The orbital characteristics stand as an item of proof that our Sun is an externally powered plasma star, instead of a sphere of hydrogen gas that is internally powered by nuclear fusion. The externally powered Sun is able to pull the Earth's climate periodically out of the Ice Age condition that has been the normal state for the last 2 million years. The 'Primer Fields' are subject to on-off conditions, depending on conditions in interstellar space.

Ice Ages result in the off periods. The on periods cause the warm interglacial climate, such as we have now, and the Dansgaard Oeschger oscillations that periodically rewarm the Earth during the glacial periods.

During the inactive state, the solar activity is reduced to a type of cosmic default level with 70% less radiated energy. At the present rate of diminishment, the solar activity phase-shift threshold to the next Ice Age period may be crossed in 30 years, or in the 2050s, most likely. With the primer system gone inactive, the climate on Earth will get 40 times colder than the Little Ice Age in the 1600s had been. Ice core evidence promises that. Without the needed preparation for human living in such an environment, 99% of humanity would die of starvation, both by the cold and by CO_2 depletion as more CO_2 becomes dissolved into the sea.

With the fields being critical for our very existence, the exploration of it is likewise critical.

In the Little Ice Age, between 10% and up to 30% of the populations in Europe had perished by starvation. The last Big Ice Age was evidently vastly harsher. Only 1-10 million people emerged from it alive. That's all we had after 2 million years of development. We want to do far better this time around; and we can, with large-scale technological infrastructures for our food supply. But will we create them? Will we get the job done in the 30 years that we still have left before the Ice Age starts anew? Will we even consider it? And how certain are we that the phase shift to the next glaciation period will begin, as the evidence suggests, in the 2050s? We have no slack on this front. Should we fail us on this absolute front, we would be committing suicide.

Numerous fields of evidence tell us that the next Ice Age is near. That's where the truth begins. Most of the evidence was discovered in the 1990s and thereafter. Some evidence is measured in ice cores; some is measured in space, by satellites. Some measurements are also made on the ground in terms of measurements of the Earth's magnetic-pole drift observed in northern Canada. All of this is seen combined with high-energy physics experiments at a leading national laboratory, and is also explored in the small in static experiments.

So, what will the answer be? Will we move with the evidence? Or will we lay ourselves down to die by default?

It takes an independent researcher to brake the taboos that have kept mainstream cosmology imprisoned, increasingly, during the past century, even while what is regarded as taboo is known to be wrong.

The Illustrated Science series is intended to open the scene beyond the threshold of accepted taboos, to where the actual physical evidence speaks for itself.

The scope of the existential challenge that the Ice Age brings with it, takes astrophysics out of the academic domain and places it into the foreground as one of the most-critical issues of our time. The big Climate Change events that have already worldwide effects are mere fringe effects in the flow of the ever-changing cosmic dynamics. The big effect, when the Ice

Age begins anew, promises to be caused by a dimmer and colder Sun. The loss of 70% of the Sun's radiated energy defines our climate future that begins in the near term.

Sure, we can live with all that by creating new platforms for agriculture that are able to operate under Ice Age conditions. But will we do it? The task is enormous. Or will we fail ourselves on this front? We have no reason to allow us to fail. We have the materials and energy resources on hand to accomplish everything that is required for us to continue to live in an Ice Age World. But will we do it? The big question that never goes away, therefore, is; will we develop our inner resources as human beings sufficiently to get the job done, and to get it done in time? Or will we do nothing, ignore the challenge, and condemn our children and one-another to an agonizing death by starvation? That's the choice.

Towards meeting the inner challenge, I have created the epic series of novels, The Lodging for the Rose. And further, towards meeting the science challenge, I have produced numerous research books and several dozen exploration videos that the Illustrated Science series is modeled after. The work is the result of a quarter century of research, for which numerous elements of evidence in related fields came to light during the timeframe of my research.

It is my hope that the work that went into all of these projects will help in some degree - for humanity that we are all a part of - to write itself a ticket to have a future.

High-resolution color images, of the images in this book, can be obtained at www.iceagetheatre.ca

*The Dansgaard Oeschger oscillations

The Dansgaard Oeschger oscillations

During the extremely weak conditions in the galaxy

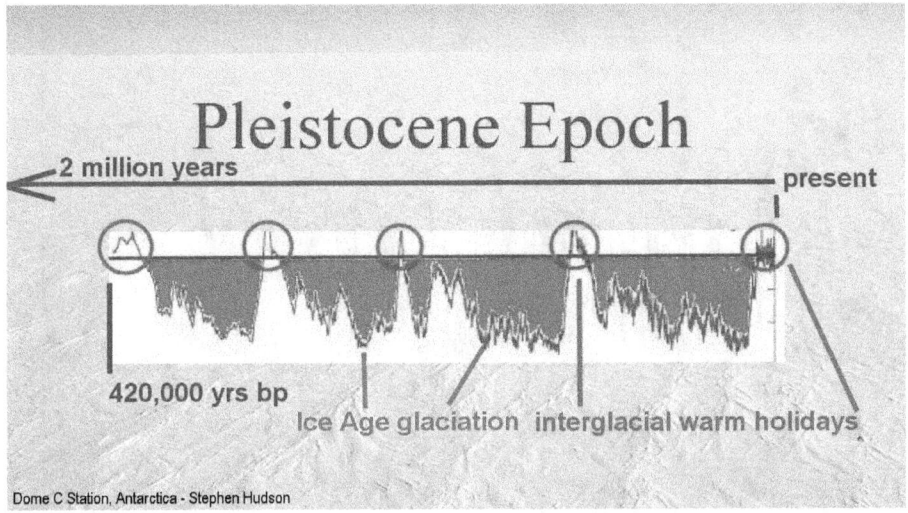

During the extremely weak conditions in the galaxy that enable the ice ages on earth, the Sun remains constantly active only during the timeframe of the interglacial pulses. In the interglacial timeframe the solar system is in its high-power mode.

During the long glacial periods

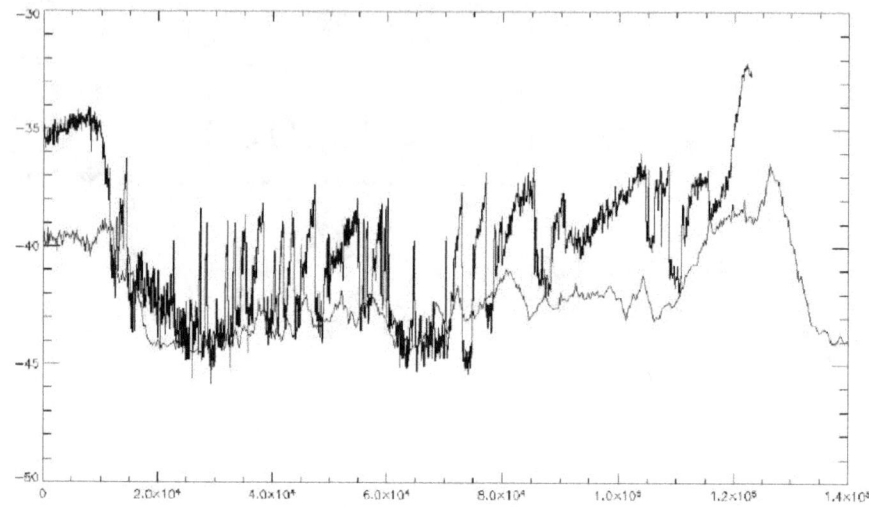

During the long glacial periods, however, between the warm and sunny interglacial pulses, large climate fluctuations have occurred that are named Dansgaard-Oeschger oscillations in honour of the discoverers of them.

These are large climate fluctuations, in the range between 20 to 40 times that of the Little Ice Age cooling. These large oscillations dominate the climate landscape all the way through the Ice Age glaciation periods. These enormously massive climate events were evidently caused by the Sun turning on and off periodically, rather than by ocean current fluctuations.

The Dansgaard-Oeschger oscillations

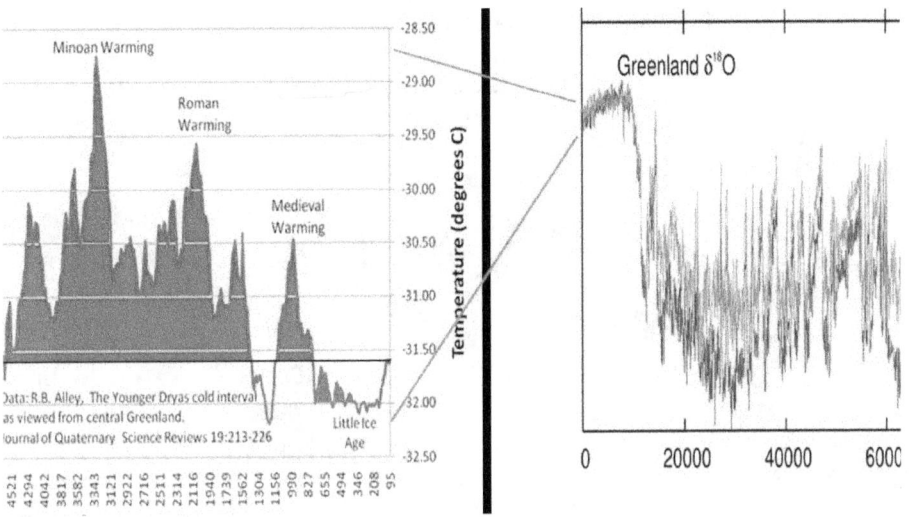

The Dansgaard-Oeschger oscillations, which have been discovered in ice core samples in Greenland, were apparently gigantic events that pale all the climate events in experienced history into insignificance.

The oscillations that have apparently spanned all the modern ice ages, may have prevented the Earth from freezing up completely, as it may have had in distant time roughly 700 million years ago, when the entire Earth became a snowball and remained frozen for tens of millions of years.

The Snowball-Earth concept is just a theory

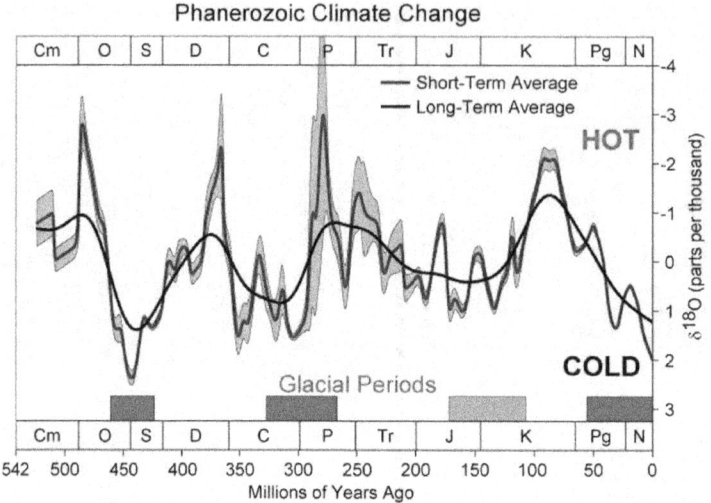

Phanerozoic Climate Change

While the Snowball-Earth concept is just a theory based on records laid up in fossils, the last four glaciation periods, that were discovered with oxygen-18 isotope measurements, coincident with data in fossils, were evidently real. It is known, for example, that the deep glaciation that occurred around 445 million years ago, was so severe at the time that half of all species of life that were known to exist at the time, became extinct. The extinction occurred not on land, but in the oceans. Life on land did not exist at the time. It evidently takes enormously massive climate events to affect the oceans so severely that half of all species living in the oceans were unable to survive. The result is referred to as the Ordovician extinction event.

The enormous rise and fall of sea-levels that occur during glaciation conditions, which had repeated exposed and flooded large areas, may have sufficiently altered the habitats to exterminate many a long-established species.

Ocean levels dropped 400 feet

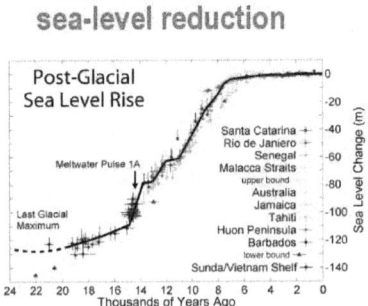

Expect a near-term sea-level reduction

The danger is that we will experience a massive reduction in sea level in the near term as the re-glaciation begins with the Ice Age Transition now in progress.
Picture a loss of only 20 meters
A powerfull new renaissance will be needed to meet the physical challenge for infrastructures

In modern time, during the last glaciation period, which is generally called the Ice Age, the ocean levels dropped 400 feet as the water became piled up on land in the form of Ice, over 10,000 feet deep, and remained frozen there. No evidence is known that the Dansgaard-Oeschger oscillations had caused sea-level fluctuations.

The Milankovitch Cycles

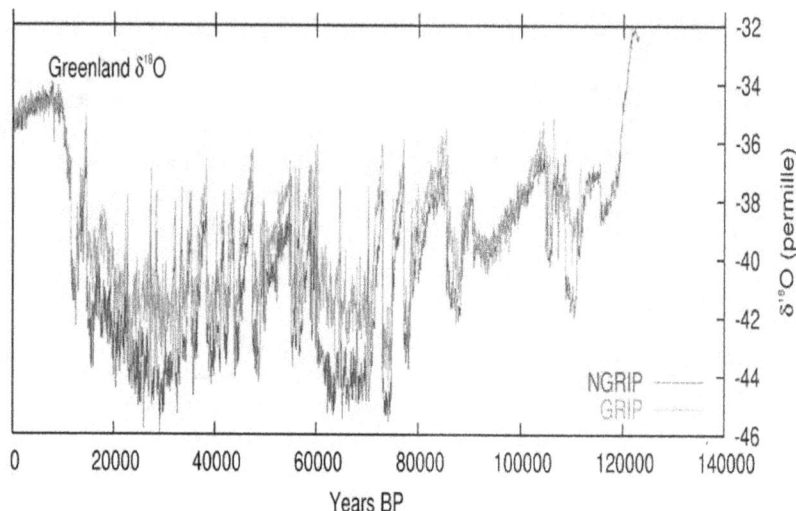

The oscillation pulses were evidently too short in duration to cause significant melting. They were primarily fluctuations between extremely cold climates and not quite as cold climates. It appears that some of the pulses came close to the interglacial level, but remained still below it. The cause for the pulses remains largely a puzzle, as their existence doesn't fit into the widely accepted Ice Age model where climate fluctuations are believed to be caused by three overlapping, long-term fluctuations of the orbit of the Earth around the Sun, which are termed the Milankovitch Cycles in honour of the man who pioneered the concept.

The Serbian geophysicist and astronomer Milutin Milankovich

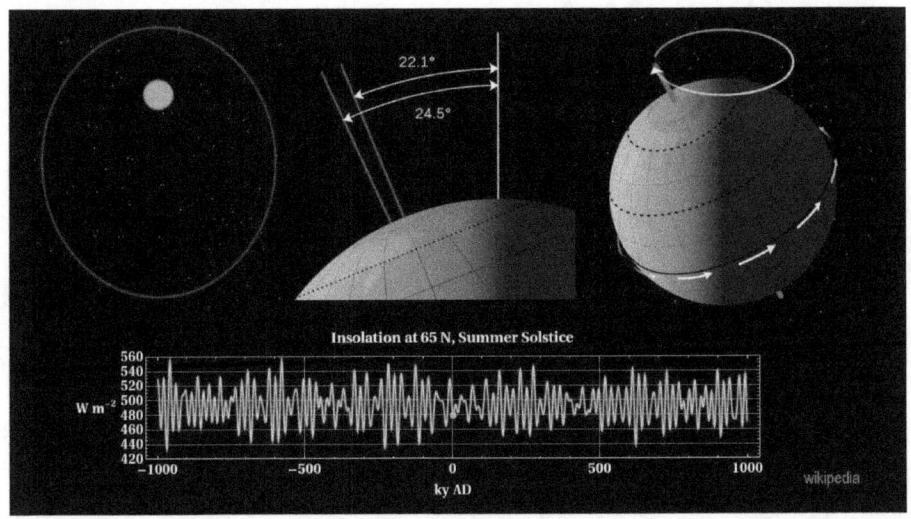

The Serbian geophysicist and astronomer Milutin Milankovich combined the 26,000-year precession cycle of the spin axis of the Earth, and the minute 41,000-year shifting of the spin-angle, with the 100,000-years cycle of the shifting eccentricity of the Earth orbit around the Sun, and theorized that the overlapping of these cycles, which cause minor variations in the seasonal and hemispheric distribution of the radiation of the Sun, causes Ice Ages to occur. This is the most elegant theory that has ever been put forth to explain the occurrence of ice ages under a constant Sun.
The problem with the theory is that the total amount of solar energy received remains always the same, no matter how the Earth's spin axis is shifting, and the eccentricity of its orbit varies.

Theatrical cause, doesn't match the measured ice core data

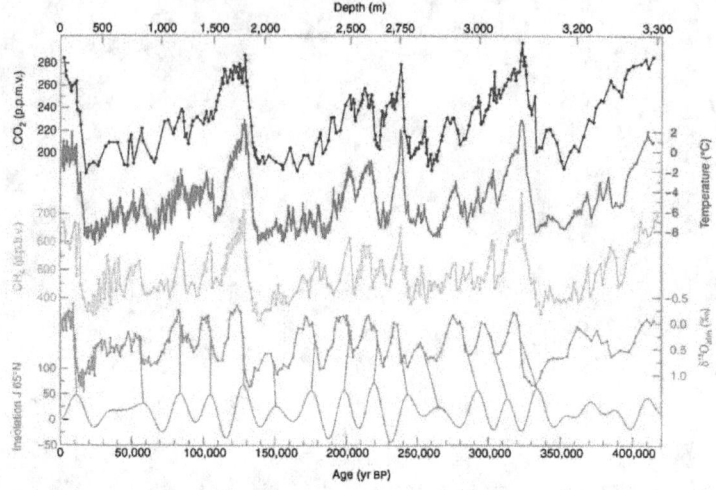

Another problem with the theory is, that the mathematically computed theatrical cause, doesn't match the measured ice core data, which renders the cycles to be subsequent phenomena of the astrophysical dynamics that cause the ice ages, rather than being causative for them. Sometimes the computed cause lags the events by 10,000 years, and sometimes precedes it by 10,000 years, as shown in the deviations presented in brown.

When it comes to the big Dansgaard-Oeschger oscillations

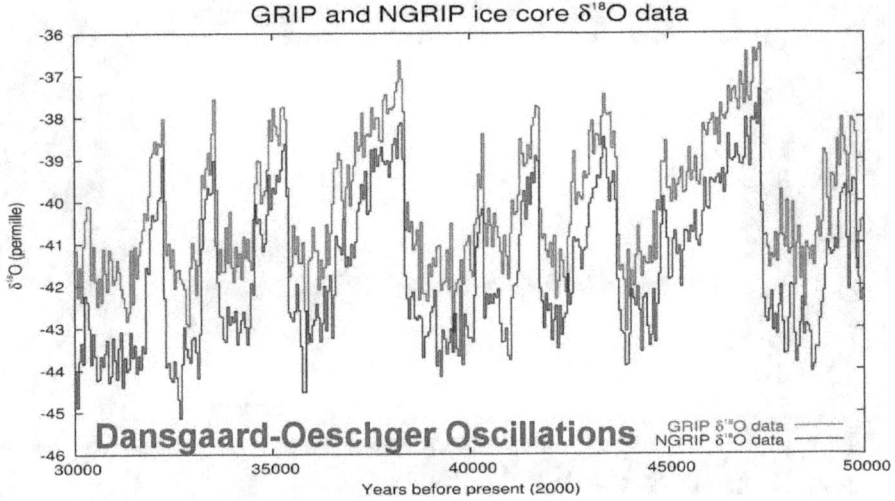

Of course, when it comes to the big Dansgaard-Oeschger oscillations that are able to spike in the range of decades and occur in intervals spaced 1470 years, the orbital cycles theory falls totally flat. A total of 25 such events have been identified in the ice core data. Unfortunately, the Dansgaard-Oeschger oscillations hadn't been discovered at the time when the Milankovich cycles theory was invented. The Dansgaard-Oeschger oscillations were only recently discovered, when the very-deep drilling through the Greenland ice sheet was undertaken.

These drilling projects are not small undertakings

These drilling projects are not small undertakings. The Greenland Ice Core Project, named GRIP, was a multinational European research project, organized through the European Science Foundation. The funding came from 8 nations, Belgium, Denmark, France, Germany, Iceland, Italy, Switzerland, and the United Kingdom, and from the European Union.

The GRIP project drilled out a 3028-metre ice core

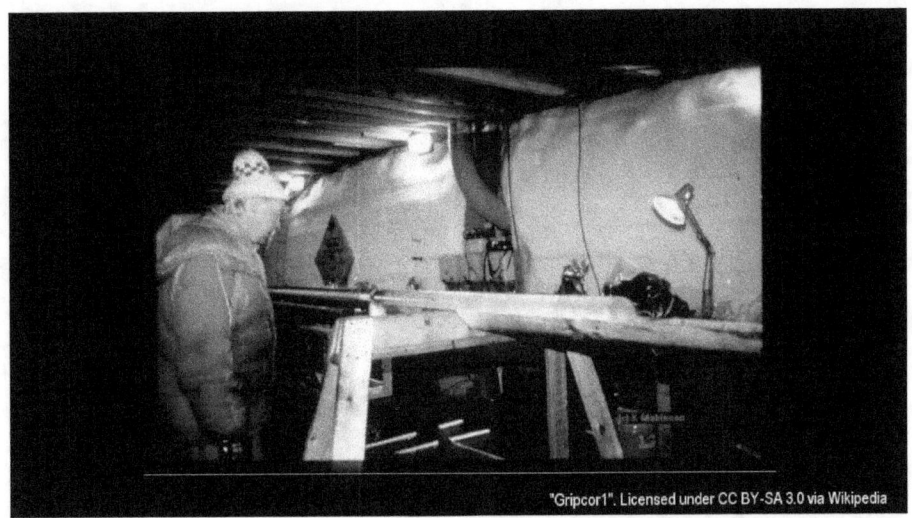

The GRIP project drilled out a 3028-metre ice core, from a summit of the ice sheet in Central Greenland, to bed rock. It took 4 years to drill this ice core, from 1989 to 1992. The oscillations were discovered in this ice core, with obviously no small surprise.

The drilling was repeated with another 4-year effort

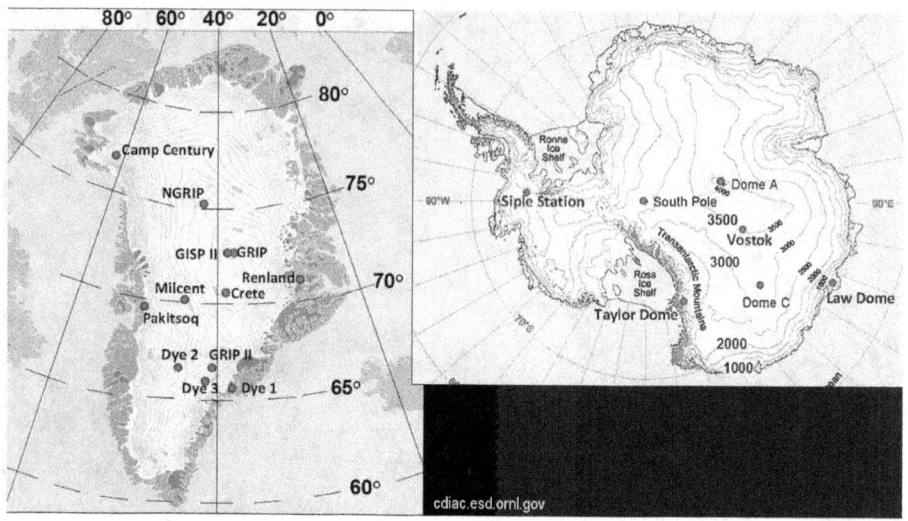

For numerous reasons the drilling was repeated with another 4-year effort. Another drilling was started a long distance further North, near the 75-degree latitude. It was started in 1999. Bedrock was reached in 2003 The drilling site and the new project was named the North Greenland Ice Core Project, or NGRIP for short. The drilling at this new site is significant in that it penetrated deeper and thereby into earlier ice, reaching back all the way into the previous interglacial period.

The new drilling confirmed

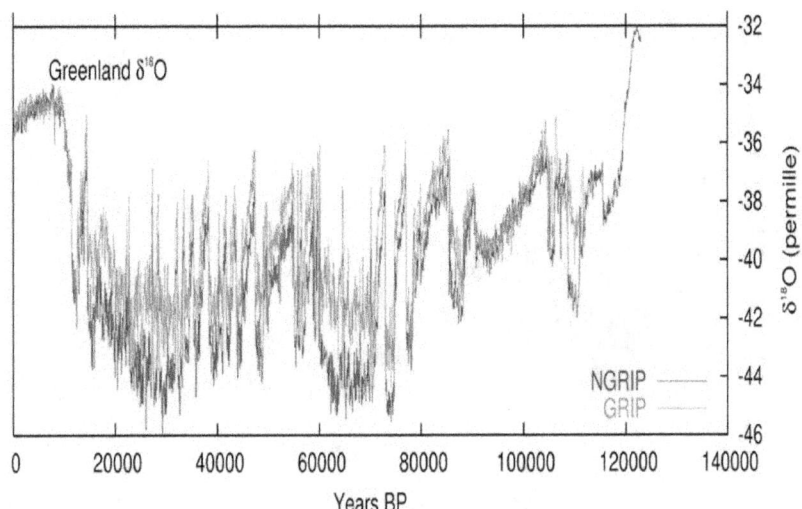

The new drilling confirmed that the Dansgaard-Oeschger oscillations were not a freak anomaly. Both the GRIP and NGRIP results are shown here, combined in different colors.

When one overlays the Greenland ice core data

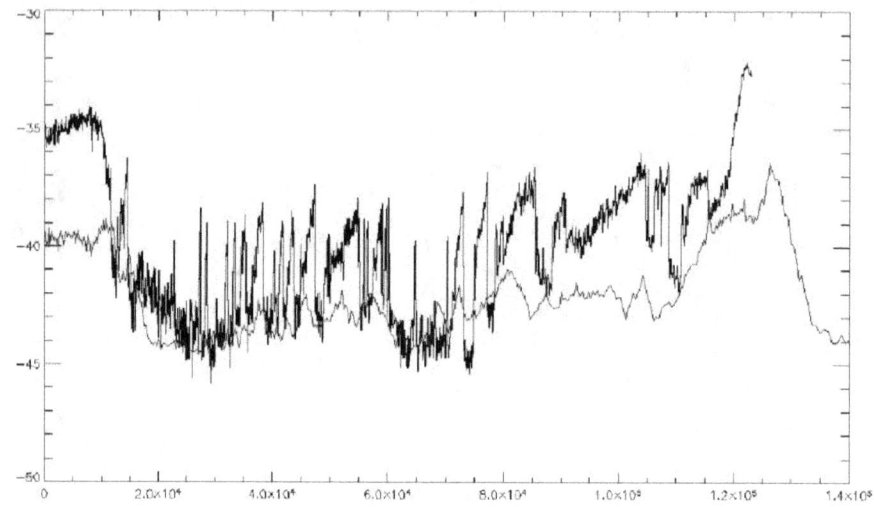

When one overlays the Greenland ice core data, with that of Antarctica, the rapid oscillations appear washed out, so that only the major trends show up in Antarctica. The reason for the difference is that Antarctica is an ice desert with extremely little precipitation whereby the fine details are lost in the coarser resolution.

The coldest, driest, and windiest continent on Earth

"Marie Byrd Land" by Michael Studinger / NASA Goddard Space Flight Center - Flickr: Marie Byrd Land. Licensed under CC BY 2.0 via Commons

It is not widely recognized that Antarctica is the coldest, driest, and windiest continent on Earth, with average 'winter' temperatures below minus 60 degrees Celsius, with the coldest temperature recorded at minus 89 degrees, that's minus 129 degrees Fahrenheit.

The snowfall there is as minuscule there as the rain in the Sahara. It may be for the lacking depth of resolution that the details in Antarctica are measured in ratios of heavy hydrogen, H-2 isotopes, while in Greenland the measurements are made in ratios of the heavy oxygen-18 isotope.

In comparison with Antarctica, Greenland is a wet place

In comparison with Antarctica, Greenland is a wet place. The same depth of ice that on Greenland covers up to 120,000 years, covers in Antarctica a range of over 450,000 years.

Ice cores go back in time up to 740,000 years

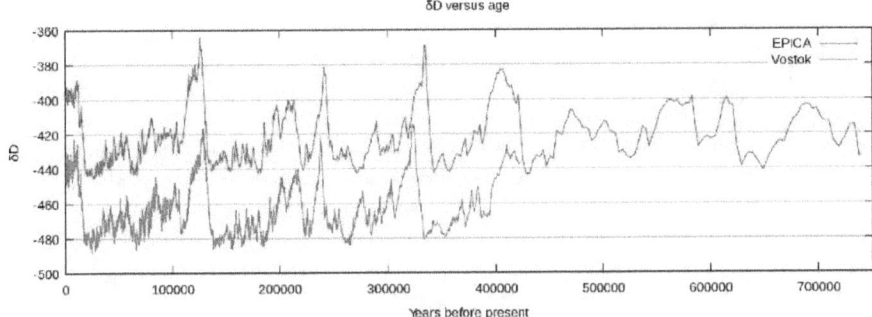

Some ice cores go back in time up to 740,000 years, as at the EPICA drilling site. The European Project for Ice Coring in Antarctica, drilled out an ice core that goes back in time 740,000 years and reveals 8 previous glaciation cycles. The project was completed in December 2004, just slightly over a decade ago.

Differences in the resolution of details

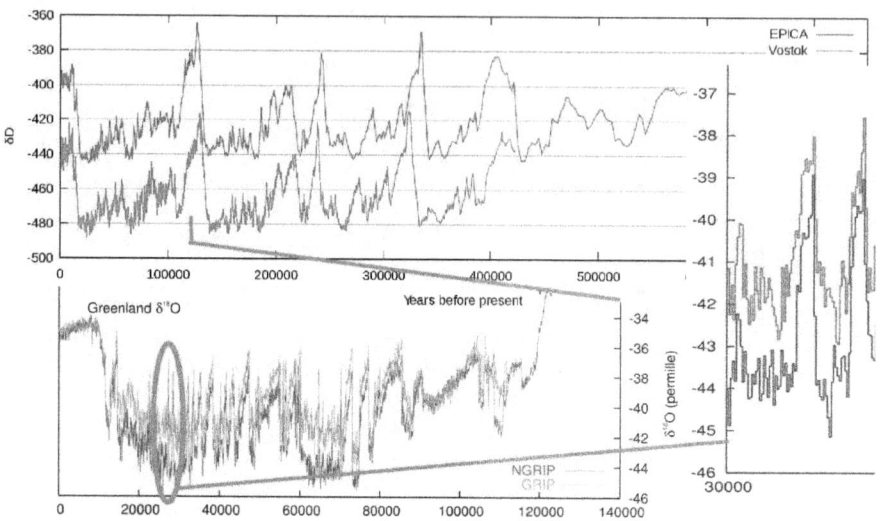

The high compaction of the ice in Antarctica all adds up to corresponding differences in the resolution of details.

The Dansgaard-Oeschger oscillations are very real

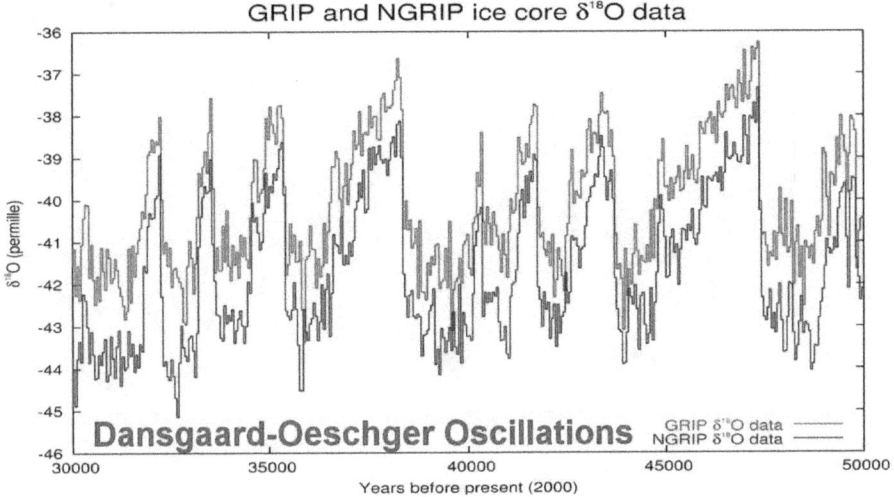

The bottom line is, as the high resolution data from Greenland from multiple drilling sites indicates, the Dansgaard-Oeschger oscillations are very real.

They are real. They are very big. And they are an enigma outside the recognition of the Sun, as being a plasma star that is able to go inactive periodically. In fact, the very existence of the oscillations proves the Sun to be an electric star, because nothing else has the enormous on-off capability that we see reflected in the historic measurements for the climate on Earth.

Spaced 1470 years on average

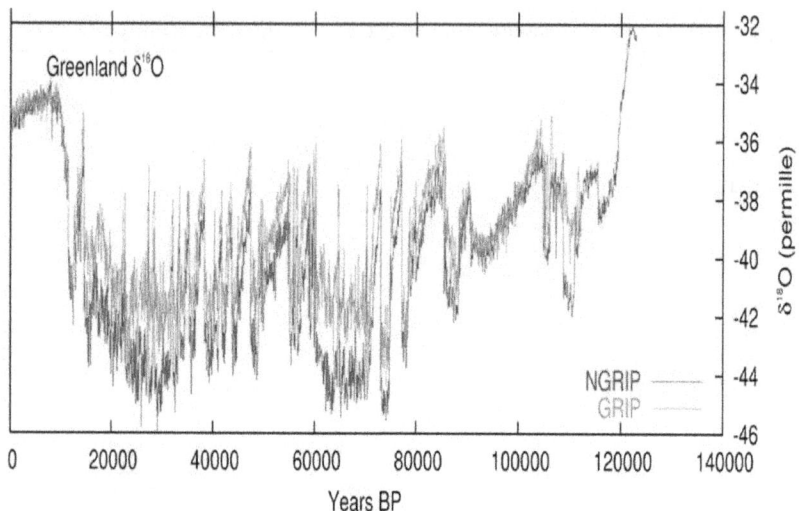

As I said before, researchers have identified 25 major Dansgaard-Oeschger pulses, spaced 1470 years on average. It might be possible that these pulses can become significant as points of reference for determining the next potential cut-off point for the Primer Fields, and with them the powered Sun.

A reasonable determination three events spaced 1470 years apart

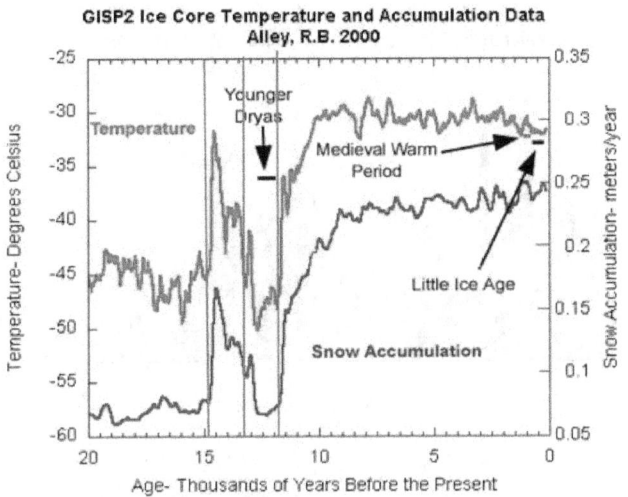

We can make a reasonable determination by looking at the last three of the Dansgaard-Oeschger events that occurred at the end of the most recent Ice Age.

We see evidence of three such events there, spaced 1470 years apart. The first started with a major upswing, followed by a down slope. The second event started on the down slope, but was of short duration. The last upswing was much larger again, and it didn't collapse this time, but unfolded further into the interglacial period.

*The most recent Dansgaard-Oeschger event

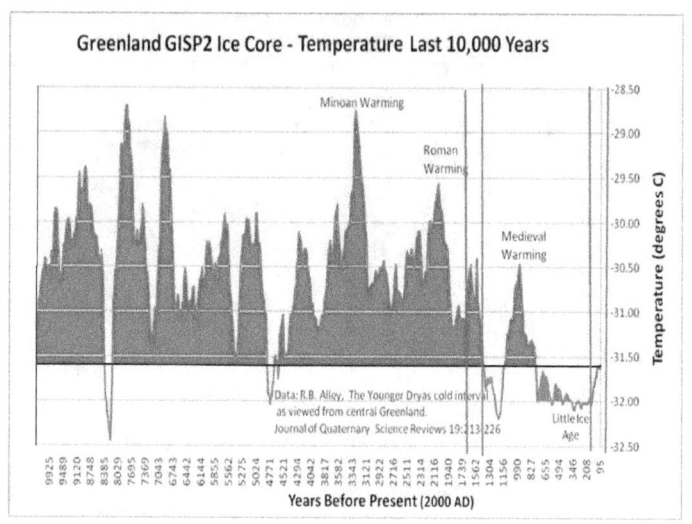

If we run the clock forward from these three steps, in steps of 1470 years, then the most recent Dansgaard-Oeschger event should have occurred roughly two to three hundred years ago. This happens to be the point in time where we see the Little Ice Age ending, and the Great Global Warming beginning.

With this coincidence in mind, a high probability exists therefore that the cosmic invigorating of the Sun and the warming of the world that broke us out of the Little Ice Age, may have been a Dansgaard-Oeschger event.

Preventing the collapse of the Little Ice Age into the next big Ice Age

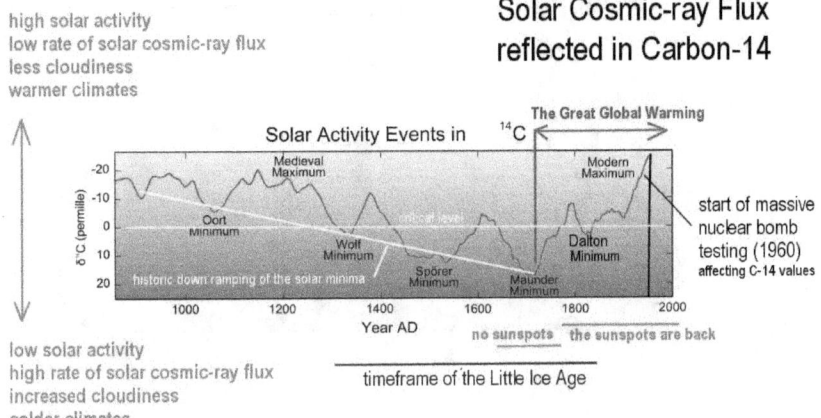

"Carbon14 with activity labels" by Leland McInnes at the English language Wikipedia. Licensed under CC BY-SA 3.0 via Commons

If this is so, the Dansgaard-Oeschger pulse at this time, becoming active in the 1700s, may have saved civilization, and humanity with it. It may have saved us by preventing the collapse of the Little Ice Age into the next big Ice Age. It overshadowed the down-ramping with a major steep up-ramping in solar activity.

33

The previous Dansgaard-Oeschger event

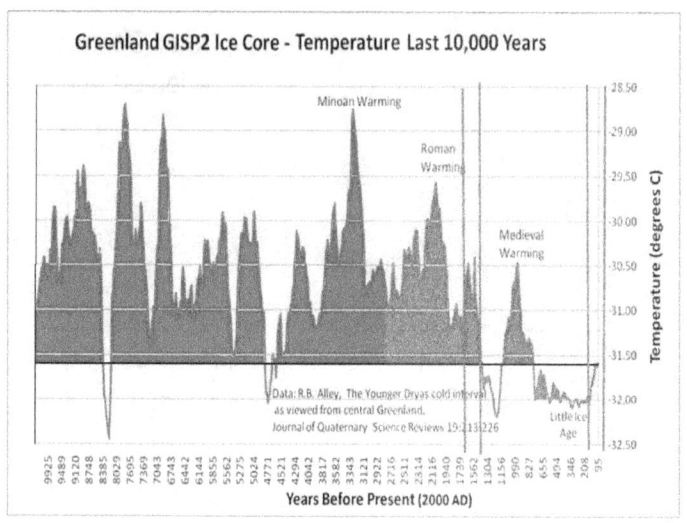

The previous Dansgaard-Oeschger event, which would have occurred 1470 years earlier, takes us back to 1770 years before the present, to the peak time between the Roman Warming, and the Medieval Warming, which had ended steeply, causing the deep low of the Oort Minimum.

If the current Dansgaard-Oeschger pulse ends in the same manner that the previous one had ended, humanity will be in deep trouble, because nothing has been prepared for the kind of world that we will then have to live in. But that's what we see unfolding, resulting from diminishing plasma density from interstellar space that we see reflected in collapsing sunspot numbers, diminishing solar activity, and diminishing solar-wind pressure, and so on.

A two-fold nested system of Primer Fields

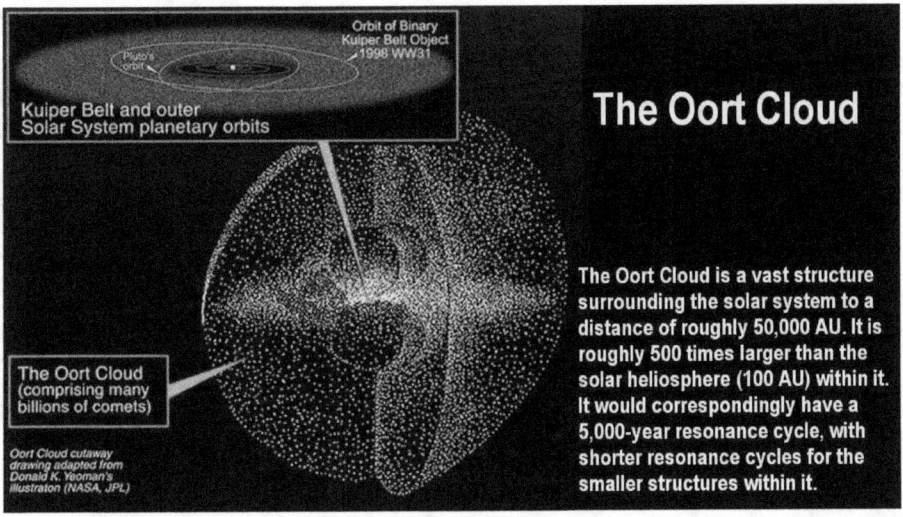

The Oort Cloud

The Oort Cloud is a vast structure surrounding the solar system to a distance of roughly 50,000 AU. It is roughly 500 times larger than the solar heliosphere (100 AU) within it. It would correspondingly have a 5,000-year resonance cycle, with shorter resonance cycles for the smaller structures within it.

The 1470-years resonance of the Dansgaard-Oeschger oscillations suggests that the electromagnetic system that focuses interstellar plasma onto the Sun is likely a two-fold nested system of Primer Fields, an inner system and an outer system. The inner system with a resonance of 22 years, would comprise the solar system and the heliosphere. The inner system, in turn, would be the focal point of the outer system that would have to be more than 50 times larger and have a 1470-years resonance. This outer system would become the dominant system during the weak time of the glaciation periods. The theorized inner Oort Cloud could meet this pre-staging requirement with a 1470-years resonance cycle. The Dansgaard-Oeschger oscillations, which are very real, may one day be seen as tangible evidence for the existence of the inner Oort Cloud. Actually, it would be surprising if this type of evidence didn't exist.

If the Oort clouds are real

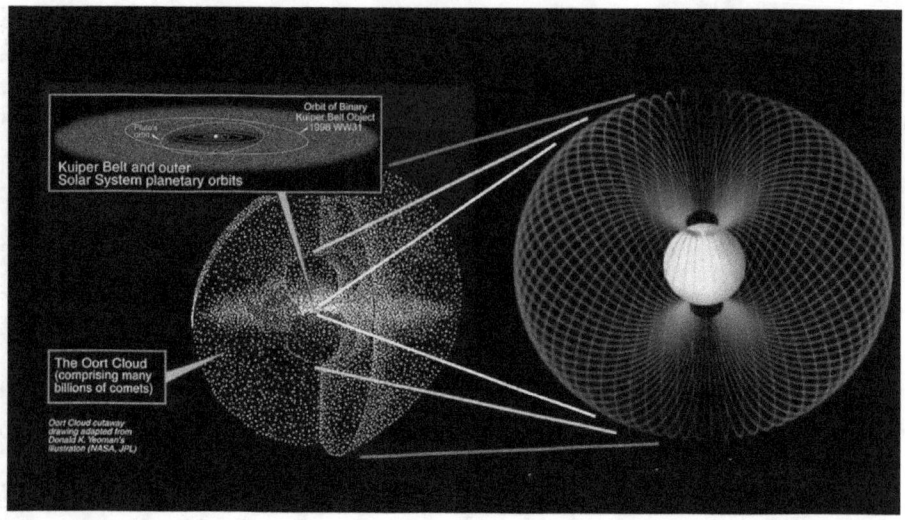

The inner cloud itself may be but the nested focal point of still another larger set of primer fields. This means that the plasma that is focused on our Sun may be supplied by a three-fold nested system of Primer Fields, of which only the innermost is subject to becoming inactive during the glaciation periods, with the resonance of the inner Oort Cloud supplying the recovery pulses that we see in the form of the Dansgaard-Oeschger oscillations in ice core samples, and so on.

If the Oort clouds are real, which they appear to be for numerous reasons, because nothing except the electromagnetic forces of the operation of primer fields would keep the space junk and comets that the clouds are made of in their distant place and centered on the Sun.

Evermore evidence keeps coming to light

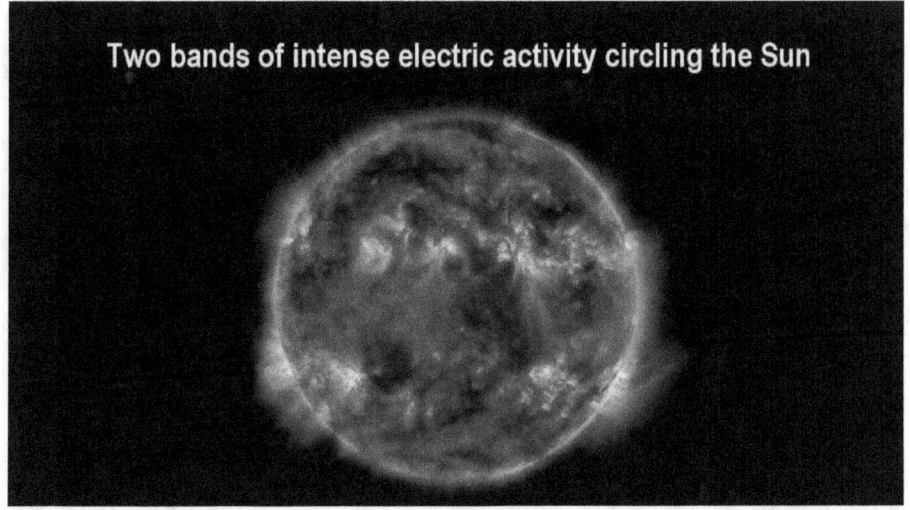

Two bands of intense electric activity circling the Sun

Evermore evidence keeps coming to light for the dynamics of the plasma powered external-fusion Sun. What we find here is ultimately our salvation, especially for the immediate times ahead. the plasma streams that power the Sun, which also extend to the Earth to some degree, may soon be required to power the human economy.

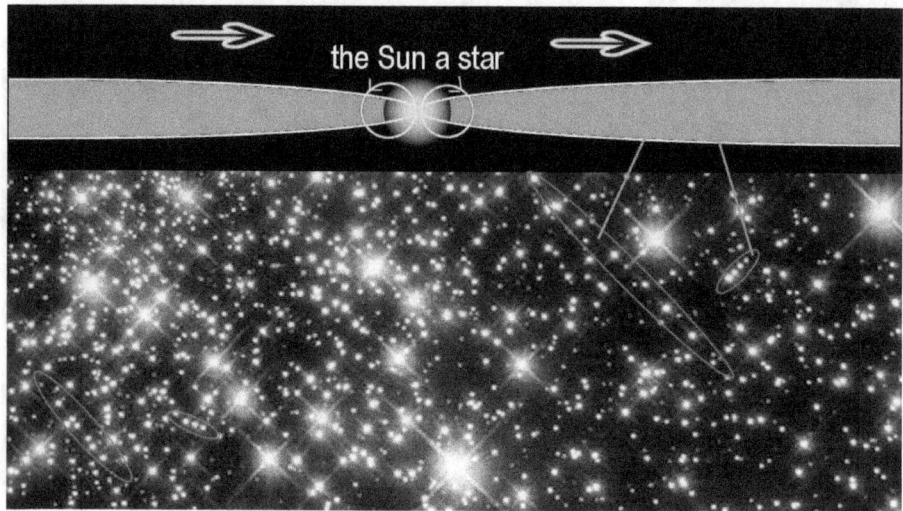

The cosmic plasma streams are not entropic. They are not subject to resource depletion. They are only subject to electric resonance fluctuations. They are inherently self-renewing from the energetic nature of space itself.

The 'Explicate Order' that David Bohm had referred to

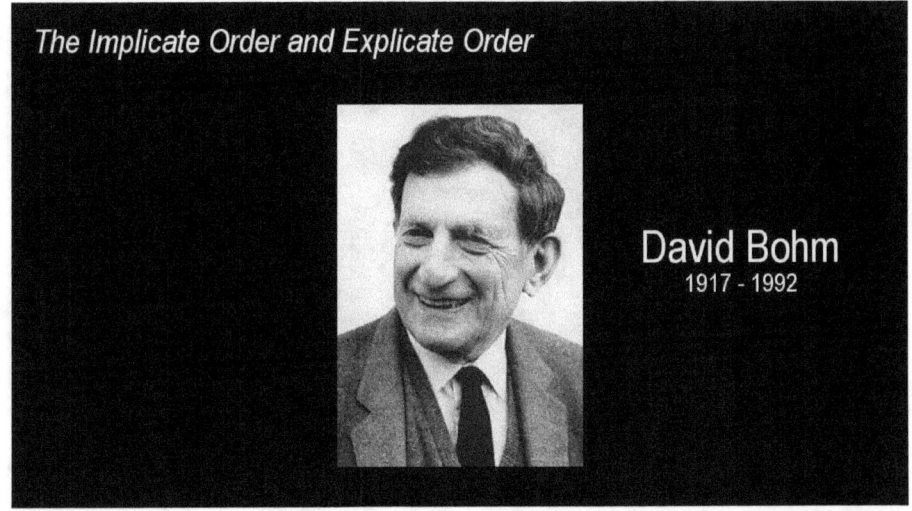

The all-pervading plasma streams appear to be unfolding along the line of the 'Explicate Order' that David Bohm had referred to, with his revolutionary concept of universal energy, that all the energies expressed in the universe are but ripples of.

Whom Einstein had once referred to as his successor

The successor

Albert Einstein (1879-1955)   David Bohm (1917-1992)

David Bohm is the man whom Einstein had once referred to as his
successor.
Whatever the case may be, plasma in space is not a fuel that is used
up, but is anti-entropic in nature as David Bohm saw the energy
background in the universe, which renders energy a dynamic part of
the universe itself. The energy background in the universe may be
the determining factor that limits the propagation speed of light
that nothing supersedes, that even the neutrinos that penetrate
everything, including the Earth itself, are bound to.

Finite resources subject to depletion

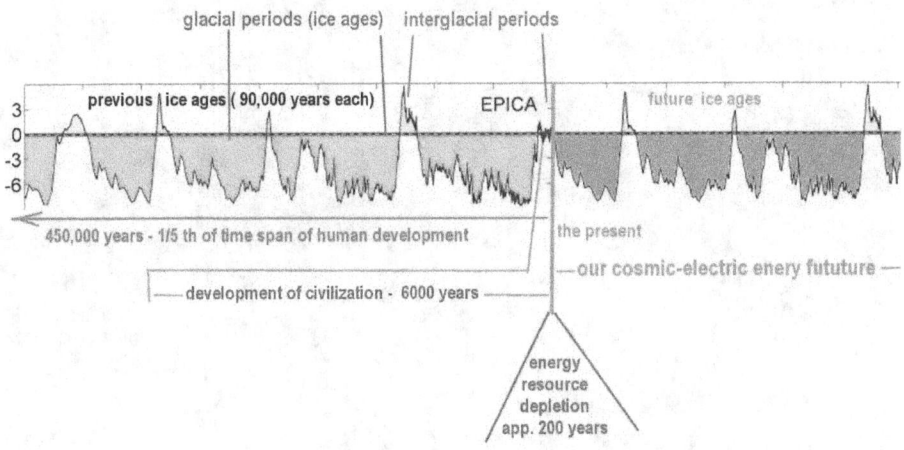

All the energy resources that are presently used to power our economies, are generated with fuels that are fast becoming depleted. Oil, gas, coal, even nuclear power fuels, are all finite resources subject to depletion. A large portion will become inaccessible when the next Ice Age begins, and those that remain accessible will be gone in a few decades, and for some types, in a few hundred years. Windmills and solar cells fall by the wayside then. All that we have remaining then to fall back on, is cosmic electric energy that may be drawn from plasma in space.

Without the utilization of cosmic energy

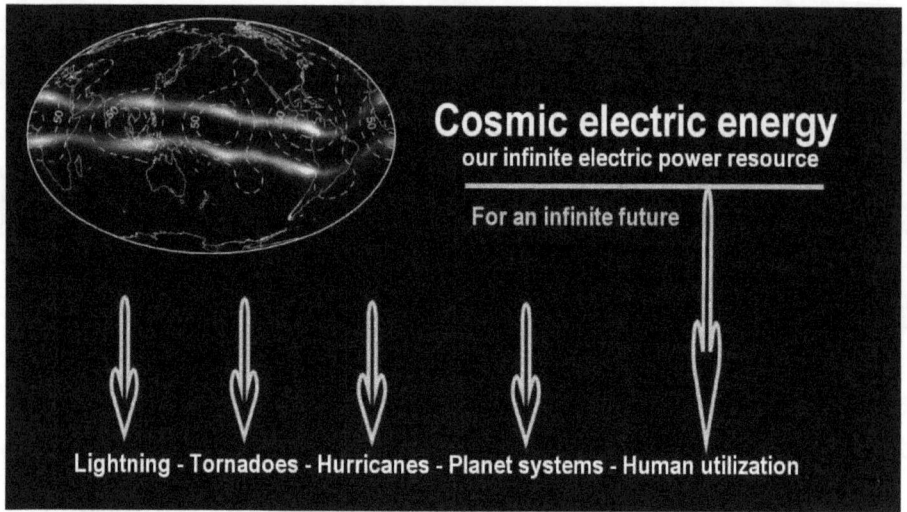

Without the utilization of cosmic energy to drive our economy, we are as good as dead. A large civilization requires large energy resources. The few recourses that we have left in the form of fuels won't last a thousand years, much less for 90,000 years till the end of the coming Ice Age, and for the millions of years thereafter that we expect to exist on the Earth. We really have no choice then, except to master the utilization of the cosmic energy system that nothing can deplete.

An interface to the cosmic power grid

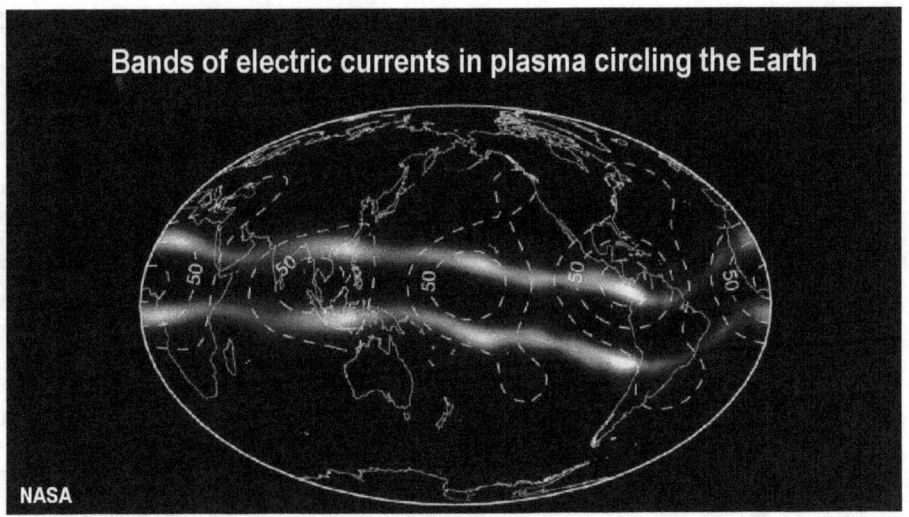

An interface to the cosmic power grid is already been made visible technologically, and exist in the form of two plasma bands encircling the Earth.

Evident on the face of the Sun in UV light

Similar bands of electric activity concentration are evident on the face of the Sun in UV light. the universe is a vast sea of energy utilization. Our galaxy of 400 billion stars is powered by cosmic energy resources, and likewise are the hundredth of billions of galaxies that exist. Energy is the 'trade name' of the universe. The universe is the product of energy. Some day we will learn that we are a part of it and open ourselves up to it. If we fail on this front we will die when the phase shift begins that will dim our Sun.

Let's stop playing those silly and dangerous games

**USAF B-2 Spirit
Stealth bomber**

20 in service
9,700 Km range without refuelling
Bomb capacity app. 38,000 pounds
Price tag: $2,000,000,000

wikipedia

So, let's stop playing those silly and dangerous games that we play, like wars, looting, depopulation, nuclear terror threats, and global warming scares. The universe has called us to attention. The time for the phase shift is getting nearer.

Soil temperatures at the Solar Terrestrial Institute in Irkutsk

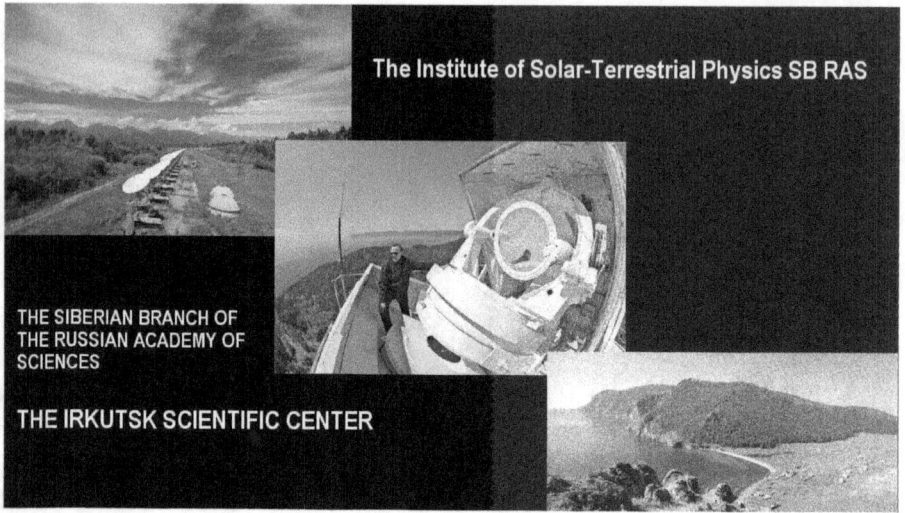

The movement towards the phase shift appears to have begun in earnest in the late 1990s. On-the-ground measurements of soil temperatures at the Solar Terrestrial Institute in Irkutsk in southern Siberia, saw a steep decline in annual average temperatures beginning in 1998 of almost two degrees over the span of 4 years. That's huge.

NASA's Ulysses spacecraft

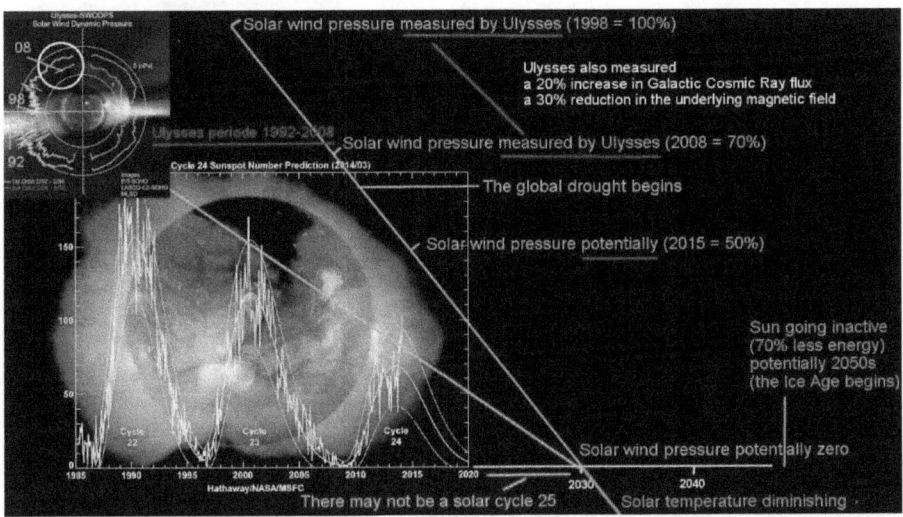

NASA's Ulysses spacecraft saw the solar-wind pressure diminishing by 30% over 10 years, likewise from 1998 on. This too, is huge for such a short time frame.

Similarly, did we see the start of a steep decline of the sunspot cycles after the 1990s, which reflect the diminishing intensity of solar activity.

Something big is in the making, and it isn't pretty.

The phase shit to the next glaciation cycle

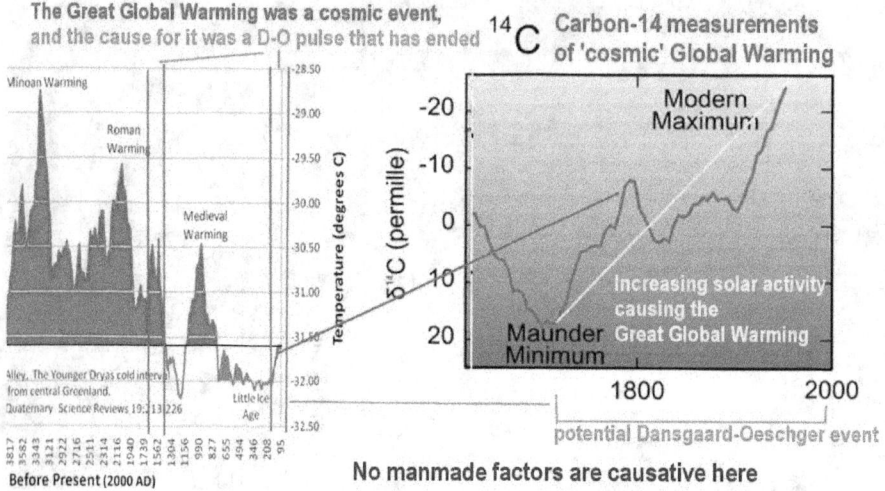

The phase shit to the next glaciation cycle even matches in time the potential end of what may have been the end of the most recent Dansgaard-Oeschger event that had pulled us out of the Little Ice Age and gave us the Great Global Warming afterwards.
And even this event is already being reversed again as the pulse that apparently caused it, is ending.

All the big indicators tell us that we are heading towards a new Ice Age in 30 years, with enormous changes unfolding for humanity.

Society likes to play fantasy games

PARIS2015
UN CLIMATE CHANGE CONFERENCE
COP21·CMP11

COP 21: Heads of delegations by GUSTAVO-CAMACHO-GONZALEZ - Licensed under CC BY 2.0 via Commons by Presidencia de la República Mexicana -delegates

Poster of the Climate Conference.
Licensed under Fair use via Wikipedia

Unfortunately, society likes to play fantasy games, such as the global warming game, instead of being concerned with what is really going on. It is betting its life, even the very existence of humanity as a whole, with just a few exceptions, that the massively indicated phase shift to the next Ice Age with a dimmer Sun will not happen, so that by this betting against all odd, nothing is being done to protect humanity's existence at its most critical stage in history. Of course, this is a bet that humanity will likely loose. The certainty stands against it.

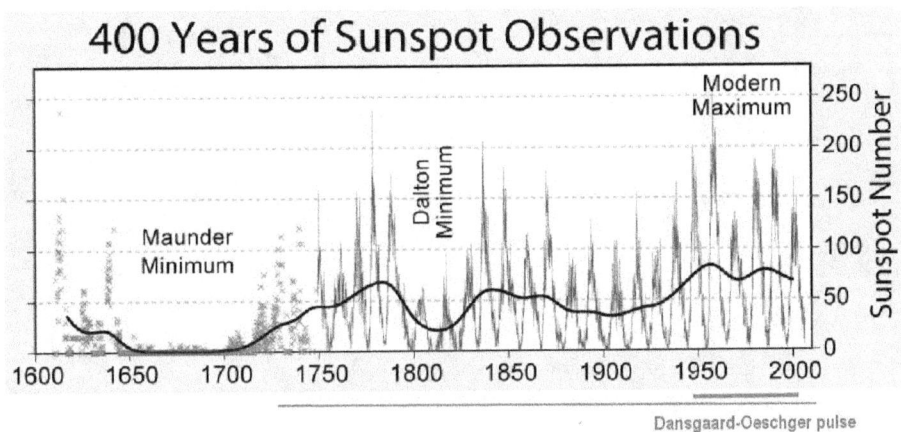

400 Years of Sunspot Observations

Dansgaard-Oeschger pulse

This means, of course, that all the global warming hoopla that was dragged onto the world scene in 1974, was essentially a fairy tale spun around the effects of what was essentially an astrophysical Dansgaard-Oeschger event. Of course, the existence of the Dansgaard-Oeschger pulses hadn't been discovered at the time, in 1974, and would not be solidly recognized until almost 20 years later.

In earlier ice cores the Dansgaard-Oeschger events were noted, roughly around 1985, as violent oscillations and were then simply dismissed and attributed to climatic anomalies. It wasn't fully apparent until the big NGRIP project was completed that these oscillations are not local anomalies, but are basic global climate oscillations.

The link of the Dansgaard-Oeschger events with the global recovery from the Little Ice Age, however, cannot be measurably confirmed. No evidence for such a link exists, other than the evidence in timing, and the fact that the recovery from the Little Ice Age with the Great Global Warming has happened and that it was a cosmic event. This

51

part is history. This should be enough, however, to move forward with, which unfortunately isn't happening. Instead, everything is stalled, and apparently intentionally so.

Give it more time to prepare

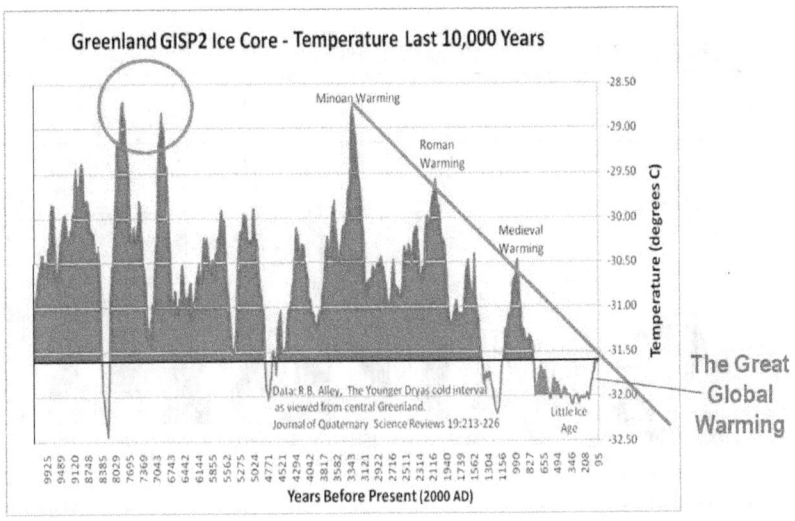

Let's hope that the currently weakening plasma environment will remain strong enough a bit longer, long enough that it will hold back the cut-off point for the Sun by a few more decades, while society learns to open its eyes. The few extra decades would give it more time to prepare the world for the dim and cold glacial environment that awaits us beyond the cut-off point of the powered Sun. Of course, this kind of pure dreaming won't get us anywhere, but in the grave.

The 25th solar cycle

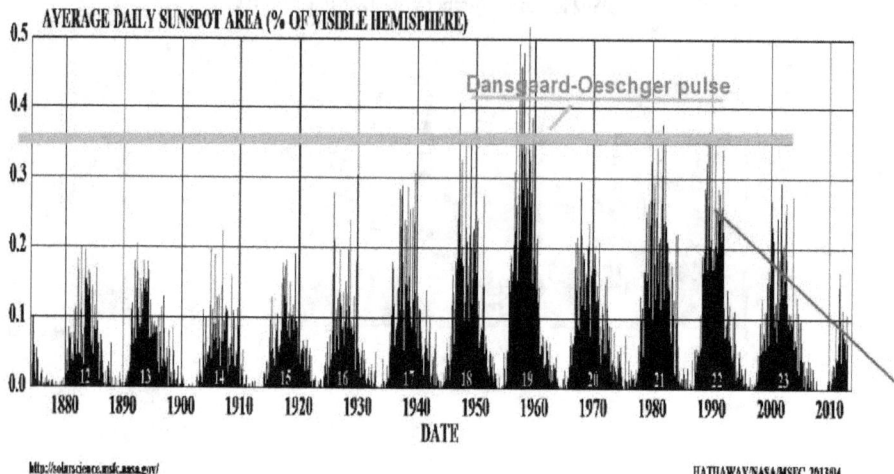

The sunspot cycles are definitely getting weaker. By the current trend, the 25th solar cycle may not have any sunspots at all. The drop-off that we see happening here, may be the ending of the current Dansgaard-Oeschger pulse that had its beginning during the Little Ice Age. It appears to have peaked from the 1950s into the 1990s, and is now dropping off fast.

The Sun is already on the track of a large weakening trend

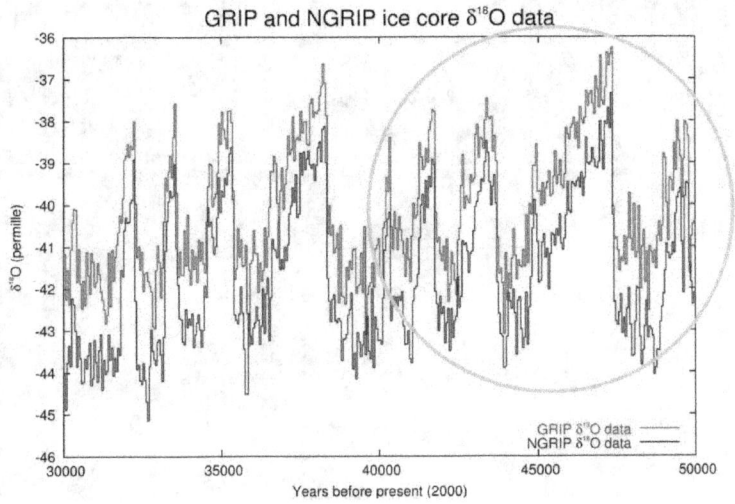

The great danger of our time is that, with the last Dansgaard-Oeschger pulse having now run its course, there is nothing much left of the interglacial plasma density to sustain the Sun, so that a free-fall collapse with the big phase shift at the end, may occur sooner instead of later and overwhelm us before we are ready for it.

The large Dansgaard-Oeschger warming events during the last glaciation period, were historically of short duration.

How close we are to the current Dansgaard-Oeschger pulse ending, which appears to be reflected in the current climate, cannot be determined. The remaining time will likely be in the range of decades, seeing that the Sun is already on the track of a large weakening trend.

What our response should be is rather obvious

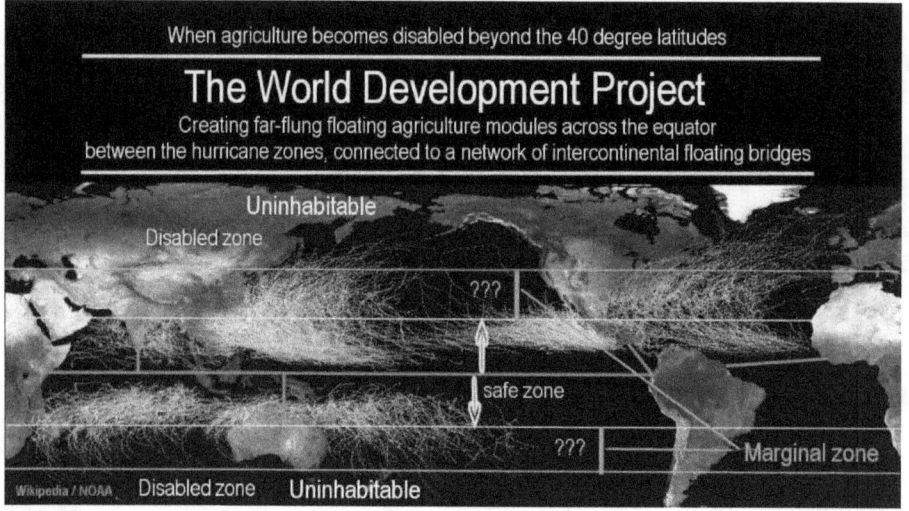

What our response should be is rather obvious. While we are committed to taking the politically correct path and do nothing, singing the global warming song that lulls us to sleep and prepares us to die in the coming Ice Age starvation, or the in the much nearer thermonuclear war that is already fully prepared for, we still have time remaining and the power on hand to secure our living with the greatest economic, cultural, and scientific development of all times that demonstrates what a human being is and is able to create with love for one-another. We have the power to choose this for our future.

Orbit dynamics of the planets

The Sun will

Evidence exists that the Sun will likely loose a portion of its mass without the action of the Primer Fields, when its inactive state begins. The external plasma pressure is thereby removed.

At a mere 71% of the velocity

Location	1. orbital period	2 mass re. Earth	3 distance (AU)	4 orbit velocity (km/sec)	5 escape velocity (km/sec)
on Sun (equator)		332,000	0		617.5
Mercury.	88 days	0.382	0.387	47.9	67.7
Venus.	225 days	0.949	0.723	35.0	49.5
Earth.	365 days	1.000	1.000	29.8	42.1
Mars.	1.88 yrs	0.533	1.520	24.1	34.1
Jupiter.	11.9 yrs	11.200	5.200	13.1	18.5
Saturn.	29.5 yrs	9.450	9.540	9.46	13.6
Uranus.	84.0 yrs	4.100	19.200	6.81	9.6
Neptune.	165.0 yrs	3.880	30.100	5.43	7.7

At the present time the planets orbit at a mere 71% of the velocity that a stable orbit requires. However, for the long glaciation periods with a lighter Sun, the present velocity would be just about right. This means that under present conditions the Sun's gravitational attraction is 29% stronger than the centrifugal force of the planets in orbit. The resulting large difference is evidently compensated for by the electromagnetic effects of the Primer Fields that maintain the orbits electrically against the stronger gravity. Evidence suggests that the orbits of the planets are electromagnetically ordered.

Non-magnetic steel balls magnetically self-spacing

David LaPoint - The Primer Fields

In a lab experiment, non-magnetic steel balls were laid on a sheet of glass. They were drawn into a specific orbit around the center of two bowl-shaped magnets. The steel balls seen here are magnetically self-spacing. If they are disturbed and let go, they return to their magnetically determined positions.

The experiment illustrates to some degree how the orbits of the planets are likely magnetically assisted, so that the Sun's gravitational variation is not a big factor.

However, during the long solar off-times in the glacial period, when the Primer Fields no longer exist, or exist only during the short periods of the Dansgaard_Oeschger warming events, the Sun's gravity would be the sole factor for keeping the orbits intact, especially in the time when the Sun's gravity is changing. While this won't likely affect the orbits of the planets significantly, it appears to have a major effect on orbiting asteroids. that are affected by cosmic drag, because of their larger surface to mass ratio.

Large dust accumulations in the ice of Antarctica

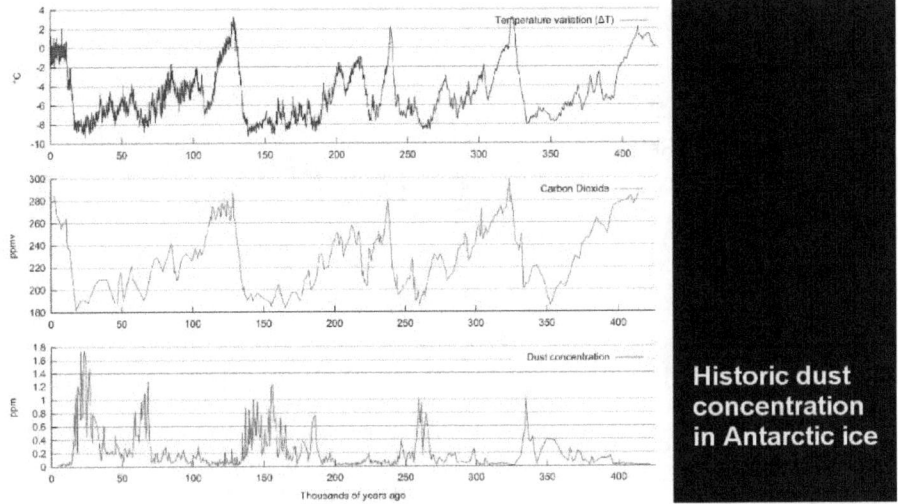

Historic dust concentration in Antarctic ice

The ice core data tells us that the loss of the primer fields has a far greater effect than we care to imagine. Near the end of each glacial period large dust accumulations are found in the ice of Antarctica, which appear to have resulted from increasing impacts of asteroids, or asteroids disintegrating in the atmosphere.

The dust always stops

Temperature Reference

Dust Accumulation

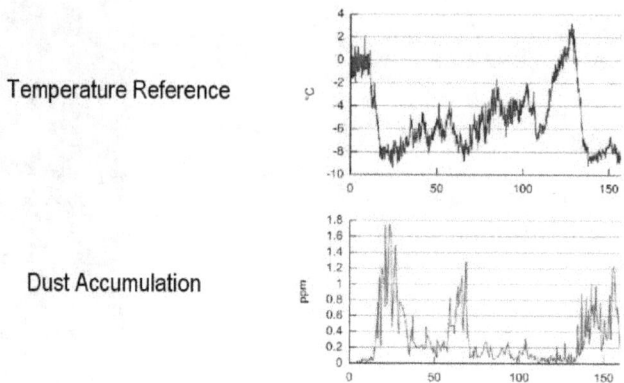

The dust always stops when the interglacial period begins in which the primer fields are active again.

*A wider field of phenomena stands as evidence for the Primer Fields

NASA

The operation of the primer fields is evidently also responsible for the planets orbiting in a tightly maintained ecliptic plain. No purely mechanistic cause would organize the planets into an ecliptic plain. No mechanistic principle would prevent the planets from orbiting the Sun in a random pattern, or force them to orbit at all. This means that the Primer Fields have a much larger effect and purpose than merely focusing plasma onto the Sun. A wider field of phenomena stands as evidence for the Primer Fields.

The near-geometrically expanding spacing

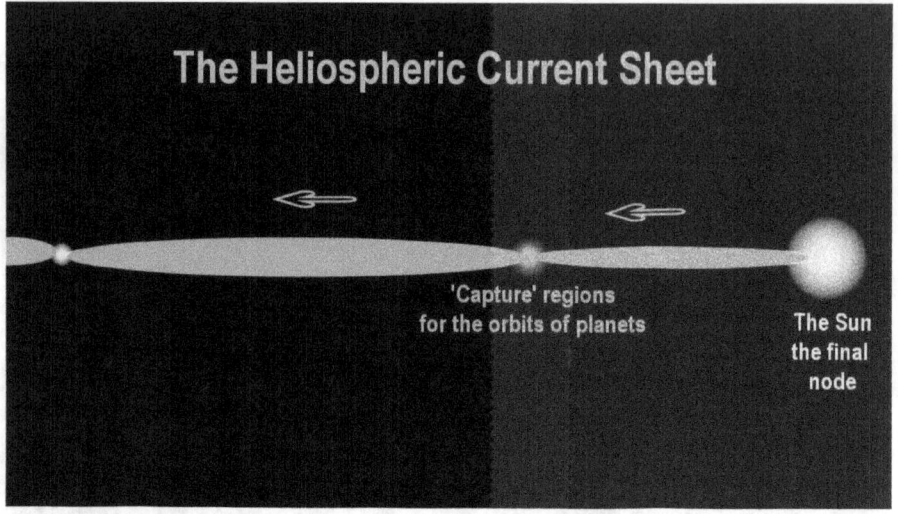

The near-geometrically expanding spacing of the orbits of the planets reflects the progressive spacing that one would expect to see between the node point in the heliospheric current sheet that becomes diffused by its expanding in geometrically widening space.

The geometrically expanding spacing of the orbits

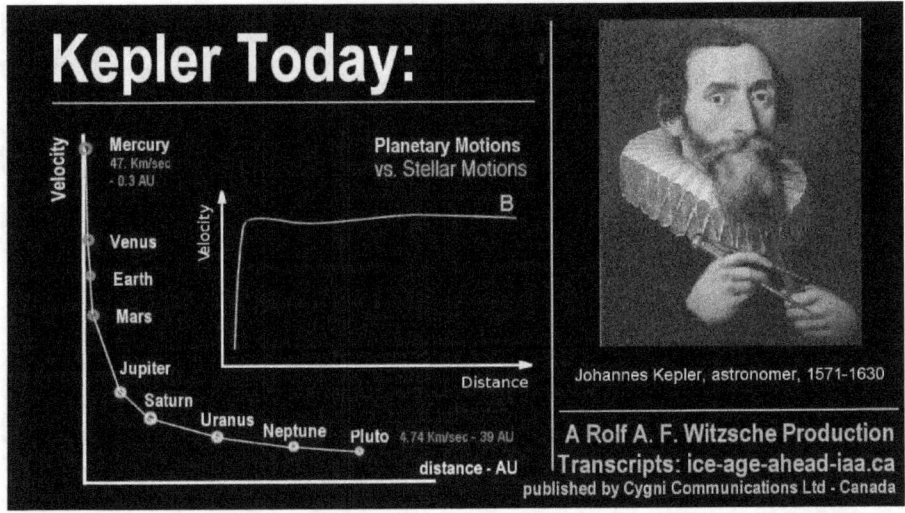

The geometrically expanding spacing of the orbits of the planets, too, adds another item of evidence for the operation, and for the far-reaching effect, of the Primer Fields that power our Sun.

The spin-rotation of the Earth itself is evidently electrically powered

wikipedia

Actually, we don't have to look quite as far for evidence. We have
mechanical evidence for the operation of the primer fields right
here on the Earth.
The spin-rotation of the Earth itself is evidently electrically powered
by the effects of the primer fields. If it wasn't for that, cosmic drag
would have stopped the spinning long ago.

Faster than the rotation of the Earth itself

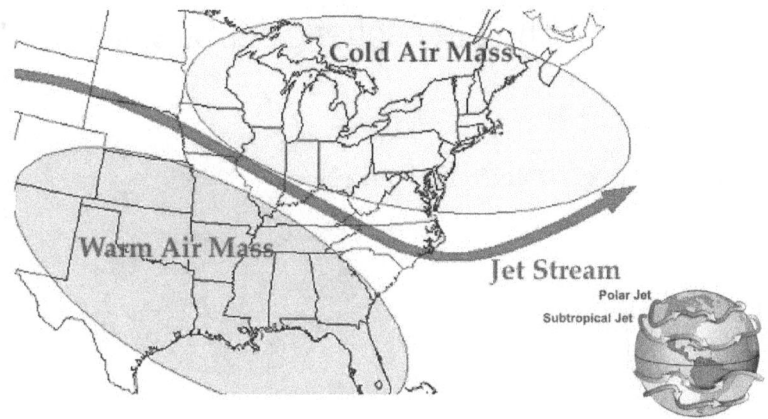

A further item of evidence that supports this recognition is found in the jet-streams in the atmosphere that are extremely fast moving air currents that flow in the direction of the Earth's rotation, but move significantly faster than the rotation of the Earth itself.

The atmospheric Jet Streams

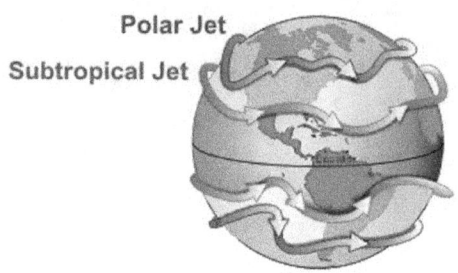

It has been long believed that the jet streams are powered by the Coriolis effect of air masses from the polar regions, reversing direction in the warmer regions. However, in this case the northern and southern jet stream would flow in opposite directions, which they don't.

*Another phenomenon where rotational movements occur faster

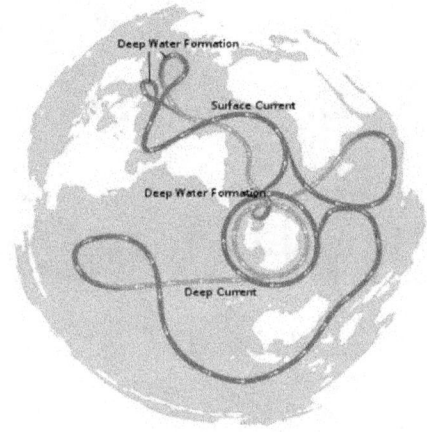

The ocean currents conveyor belt centered on the deep cold waters encircling Antarctica

"Conveyor belt" by Avsa - under CC BY-SA 3.0 via Commons -

Another phenomenon on Earth where rotational movements occur faster than the rotation of the Earth, are the massive ocean currents that flow in a circle around Antarctica.
The currents shown here are a part of a worldwide cold-water recycling system.
It appears to be thermally powered. But it also has the potential to be electrically powered, which is probably more likely.

Cold waters in the polar regions The Arctic pool has an outflow

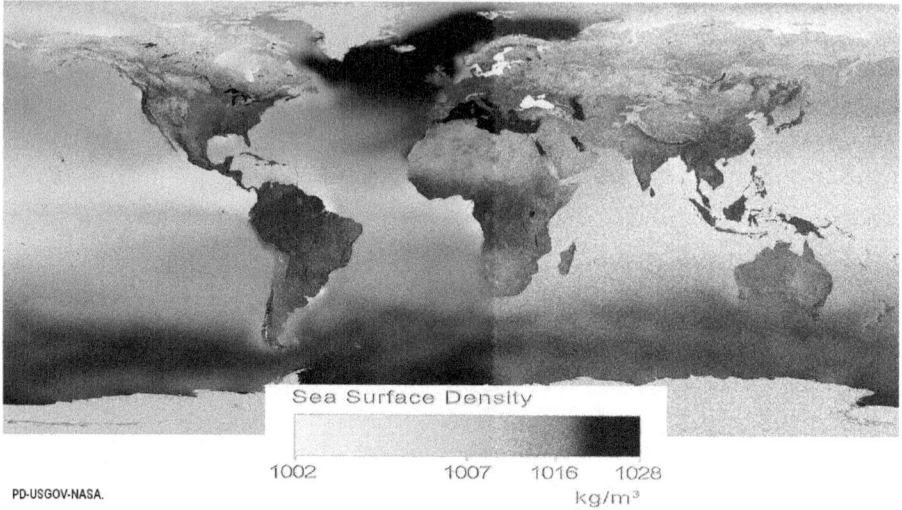

Cold waters in the polar regions are so dense by contraction and salination, that they sink into deep pools.

The Arctic pool has an outflow that creates a deep current that flows all the way to Antarctica where it joins the deep pool there, that encircles the continent.

Two currents 'spin' off from the rotating pool

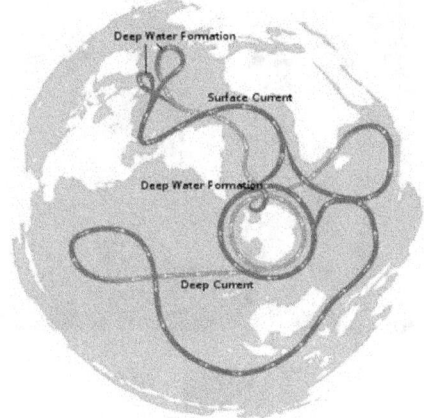

The ocean currents conveyor belt centered on the deep cold waters encircling Antarctica

"Conveyor belt" by Avsa - under CC BY-SA 3.0 via Commons -

With the encircling deep pool flowing faster than the rotation of the Earth, two currents 'spin' off from the rotating pool. One flows to the coast of Africa and into the Indian Ocean where the cold waters warm up and resurface. The second branch flows into the central Pacific, where it too warms up and resurfaces.

The cold deep waters originating in the polar regions

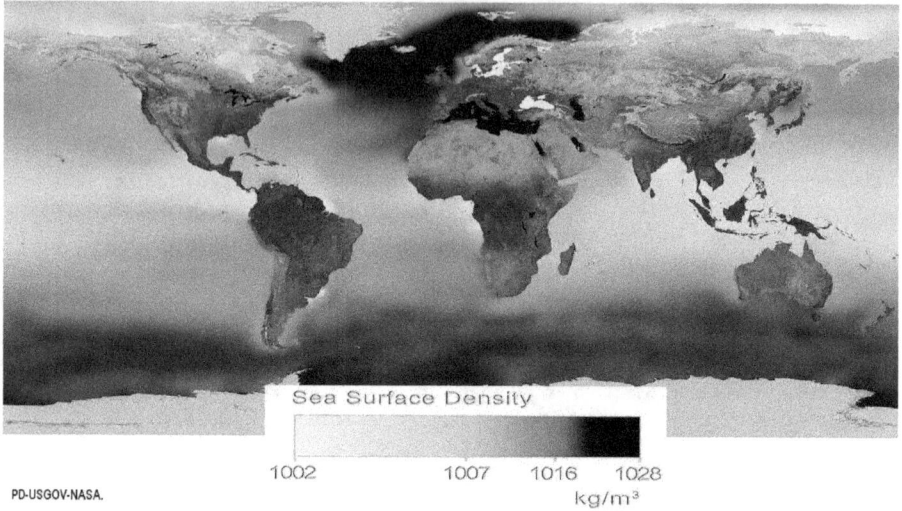

Sea Surface Density

1002   1007   1016   1028

kg/m^3

With the cold deep waters originating in the polar regions, they carry with them large volumes of dissolved CO_2. Cold waters can dissolve CO_2, to up to 10 times greater density than the density than the atmosphere presently holds. In this manner, dissolved CO_2 takes a ride in a very slow moving recycling system with a transit time of 350 years to a thousand years or more, which is evidently electrically powered by rotational actions motivated by the primer fields.

Recycling system emits CO2 dissolved 350 years ago

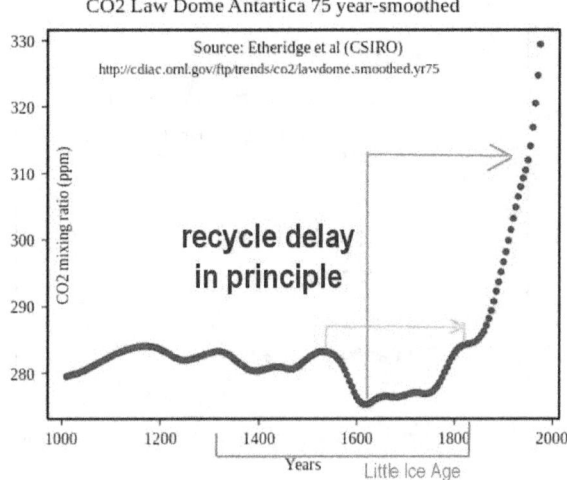

This means that the actively-powered recycling system emits CO2 into the atmosphere today that had been dissolved into the oceans during the extreme cold time of the Little Ice Age. This also means that the large CO2 increase in modern time originated at the Little Ice Age 350 years ago, or in other cold periods in more distant times, rather than being the product of human activity.

The recycling system incorporates three different transit times

CO2 Law Dome Antartica 75 year-smoothed

If one considers that the recycling system incorporates three different transit times, it is possible for CO2 to be released by the recycling system from 3 different historic times simultaneously.

The transit time to the Indian Ocean

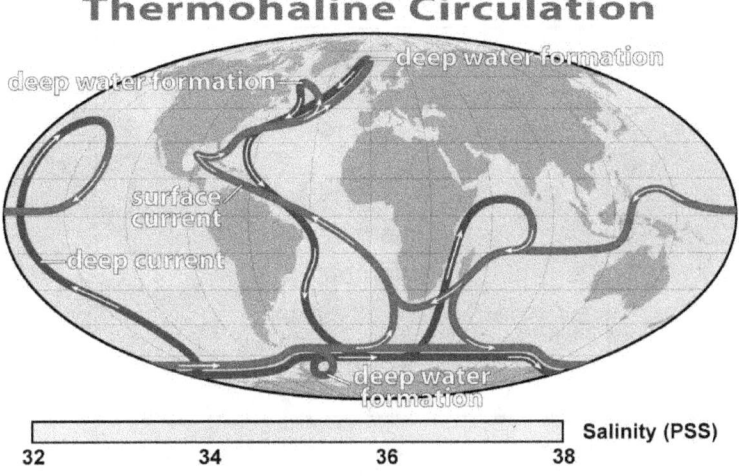

The transit time to the Indian Ocean is the shortest, possibly in the range of 350 years, with the island of Madagascar in the way. The transit time to the Mid Pacific appears to be significantly longer. And for the CO2-rich waters from the Arctic in the North, the total transit time, probably exceeds a thousand years. The exact times are not known.

Different transit times resurface CO2 from three different eras

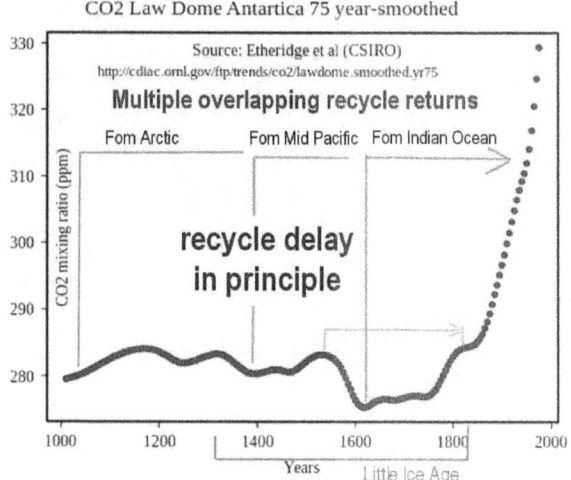

CO2 Law Dome Antartica 75 year-smoothed

Since the different transit times resurface CO2 from three different eras simultaneously, the large increase in CO2 that has been measured in recent times appears to be the result of overlapping recycling returns from three of the big cold periods of the last twelve hundred years.

The large increase in CO2 that has occurred

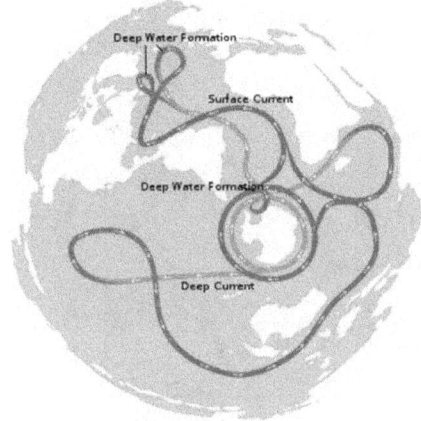

The ocean currents
conveyor belt
centered on the
deep cold waters
encircling Antarctica

"Conveyor belt" by Avsa - under CC BY-SA 3.0 via Commons -

This means that the large increase in CO2 that has occurred, which has invigorated the biological system, is a benefit provided for humanity by the electromagnetic effects of the Primer Fields that evidently power the recycling system by the rotation of ocean currents around Antarctica at a speed greater than the rotation of the Earth. The increased CO2 is a benefit for us, for the simple reason that CO2 is critical for plant growth and thereby increases harvests.

The Great Global Warming resulted from increase in solar activity

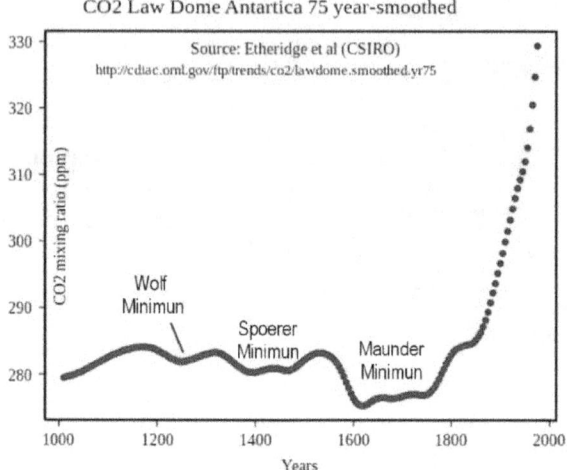

CO2 Law Dome Antartica 75 year-smoothed

Of course other factors also enter the consideration of the CO2 increase that the measurements tell us of. One factor is, that the increase in CO2 that is shown here, has occurred during the timeframe of the Great Global Warming, which has measurably resulted from the sharp increase in solar activity that has occurred during this timeframe from the Maunder Minimum on.

The Great Global Warming that has recovered the global climate

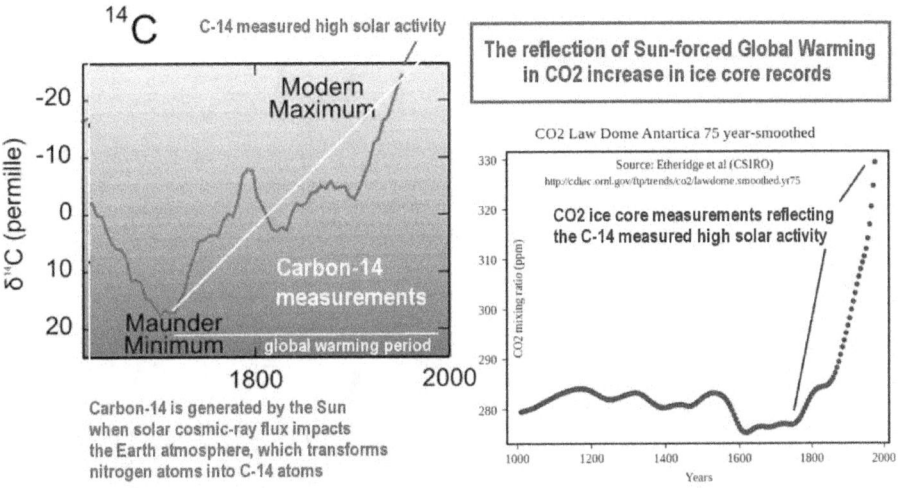

The Great Global Warming that has recovered the global climate from the Little Ice Age, was clearly the result of the sharp increase in solar activity that occurred after the Maunder Minimum. The sharp increase in solar activity has been recorded in increased sunspot cycles, and also in Carbon-14 measurements, which measure solar cosmic-ray flux. The resulting warm climate from increased solar activity would naturally increase the atmospheric CO_2 density, as less CO_2 is being dissolved into the oceans during the warm periods.

Topics to think about:

* The reflection of Sun-forced Global Warming in CO_2 increase in ice core records

* Carbon-14 is generated by the Sun when solar cosmic-ray flux impacts the Earth atmosphere, which transforms nitrogen atoms into C-14 atoms.

* The C-14 measured, high solar activity

*CO_2 ice core measurements reflecting the C-14 measured high solar activity

* Solar-forced global warming period

*The measured 15% increase in CO2 from the 1950s on

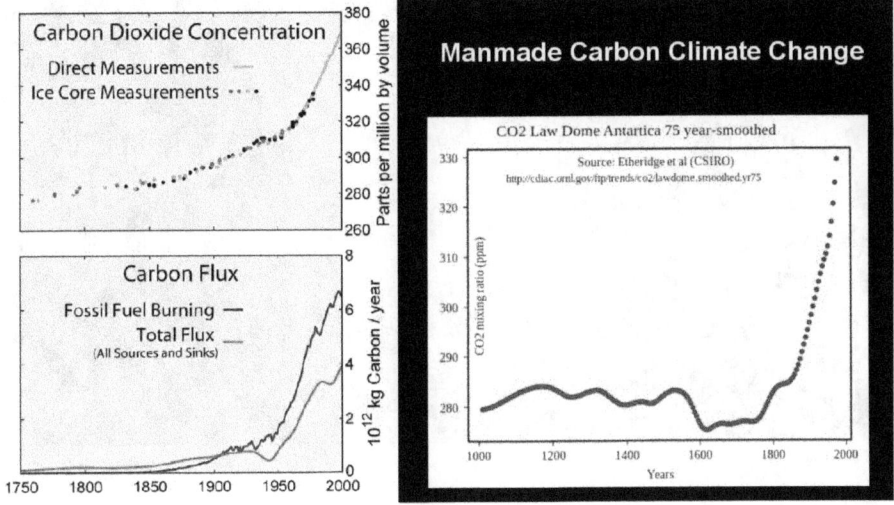

It is interesting to note that the measured 15% increase in CO2 from the 1950s on, couldn't have originated from manmade contribution that, at the peak of the increase, amounted in total a mere 1% of the global CO2 per year.

A quarter of the global atmospheric CO2 gets recycled annually

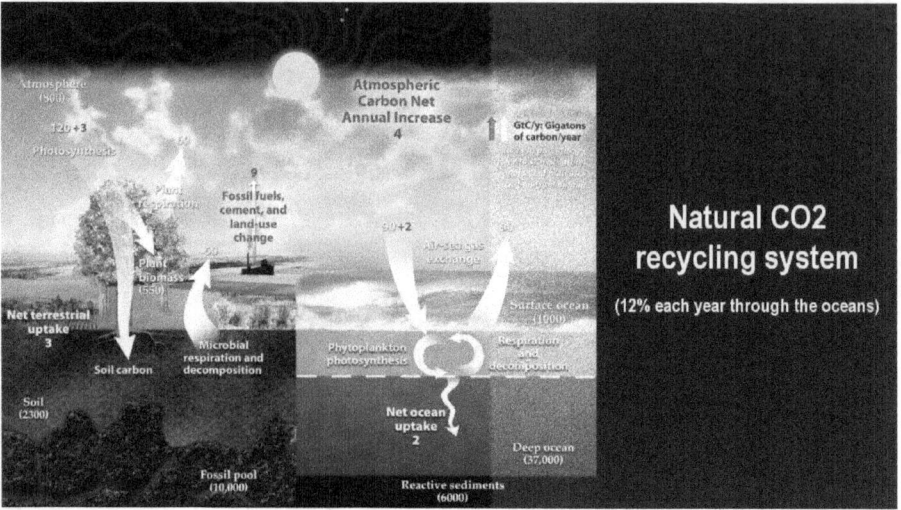

While a quarter of the global atmospheric CO2 gets recycled annually, with slightly half of that flowing through the ocean recycling system.

It is further interesting to note

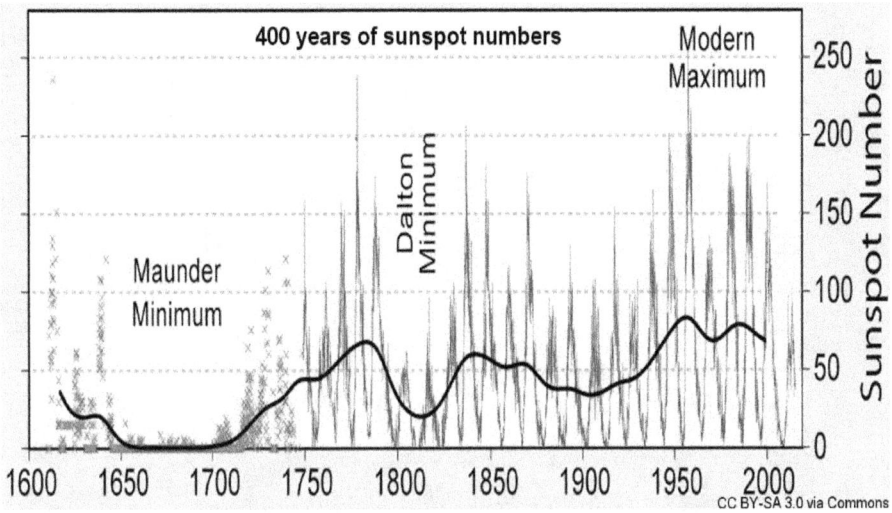

400 years of sunspot numbers

Modern Maximum

Dalton Minimum

Maunder Minimum

Sunspot Number

CC BY-SA 3.0 via Commons

It is further interesting to note that the period from the 1950s on to slightly before 2000 was a peak period in solar activity, and the warmest in hundreds of years, which should therefore coincide with a high rate of CO2 increase due to the warming.

The sharp CO2 increase that the ice core data indicates

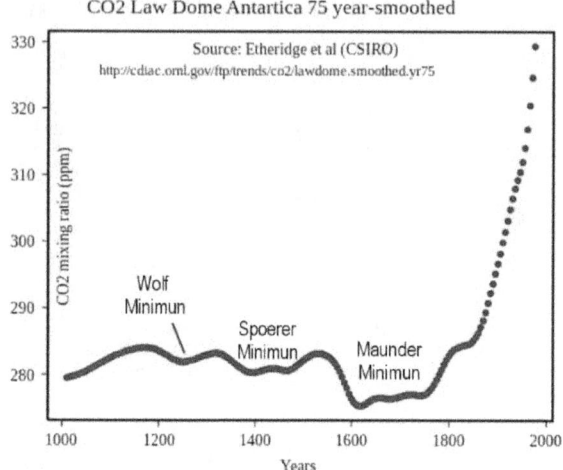

In addition, the sharp CO2 increase that the ice core data indicates, has been disputed by a high-level expert in the field.

Dr. Zbigniew Jaworowski,

Zbigniew Jaworowski, PhD, MD, D.Sc.
chairman of the Scientific Council
of the Central Laboratory
for Radiological Protection in Warsaw

"CO2 - the greatest scientific scandal of our time"
for details, see: ice-age-ahead-iaa.ca

... a world-renowned atmospheric scientist
and mountaineer, who has excavated ice
out of 17 glaciers on 6 continents in his
50-year career.

21st Century Science and Technology

Dr. Zbigniew Jaworowski, the one-time chairman of the Scientific
Council of the Central Laboratory for Radiological Protection in
Warsaw, has pointed out that ice core measurements have an
inherent problem.

With CO2 being highly soluble in water

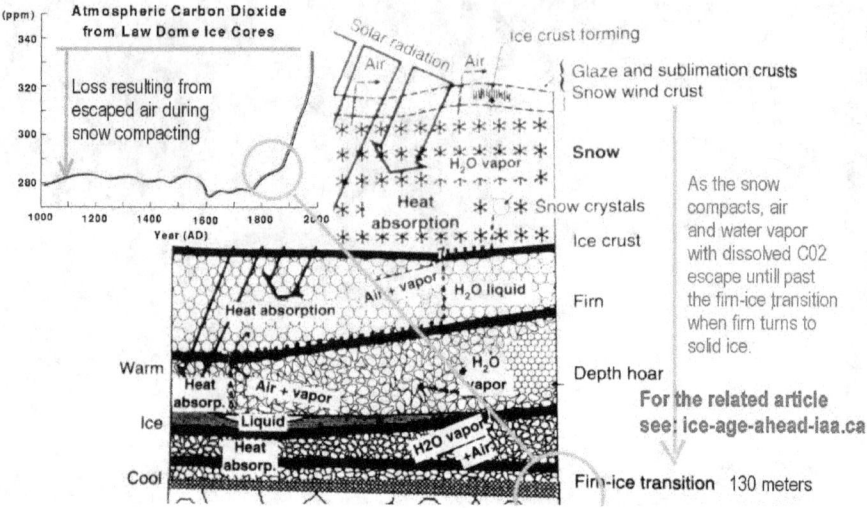

His experience has been, resulting from a 50-year career of ice explorations on 6 continents, that large losses of CO2 result in the period of snow becoming compacted, first into firn, then into deep hoar, and so on, until it becomes compressed into solid ice. With CO2 being highly soluble in water, a large portion of CO2 in the original snow, becomes lost with the loss of water vapor that is expelled by the compaction process. His take is that the CO2 portion that is finally compressed into ice, is significantly less than what was originally present.

Dr. Jaworowski compared the CO2 concentration in ice

Zbigniew Jaworowski, PhD, MD, D.Sc.
chairman of the Scientific Council
of the Central Laboratory
for Radiological Protection in Warsaw

"CO2 - the greatest scientific scandal of our time"
for details, see: ice-age-ahead-iaa.ca

... a world-renowned atmospheric scientist
and mountaineer, who has excavated ice
out of 17 glaciers on 6 continents in his
50-year career.

21st Century Science and Technology

Dr. Jaworowski compared the CO2 concentration in ice from the stage just past the firn-to-ice transition, with CO2 concentrations of the same time in deep sediments, and reports that the CO2 level in the sediments hadn't significantly changed in this period.

Dr. Jaworowski notes that the hokey-stick phenomenon

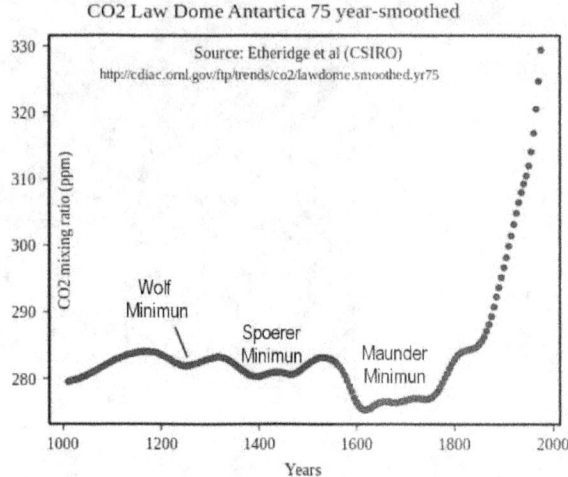

Dr. Jaworowski notes that the hokey-stick phenomenon that has been deployed to scare humanity into economic suicide measures and depopulation, is not actually true, but is simply a characteristic of the CO2 loss in the ice compaction process. He notes that the characteristic is known, but appears to be intentionally abused for political purposes.

In this sense the Primer fields even affect politics

PARIS2015
UN CLIMATE CHANGE CONFERENCE
COP21·CMP11

Poster of the Climate Conference.
Licensed under Fair use via Wikipedia

COP 21: Heads of delegations by GUSTAVO-CAMACHO-GONZALEZ - Licensed under CC BY 2.0 via Commons
by Presidencia de la República Mexicana -delegates

In this sense the Primer fields even affect politics. Under political dogmas the imagined CO_2 increase in modern time is regarded as man-made, and is deemed to be the sole cause for the Great Global Warming that occurred after the Little Ice Age when industrial activity begun, which, as it is said, must be prevented with political will, at all cost, from further increasing. Not a word was ever spoken in these arenas that the Great Global Warming was demonstrably caused by the Sun, and was impelled by cosmic processes and resonating astrophysical conditions, while CO_2 is not a climate factor at all in comparison.

Nor was it ever noted in these politically-driven operations that the hockey-stick CO_2-increase is an illusion caused by the false interpretation of the ice core data. Nor was it ever said that the actually, physically measured increase in CO_2 levels, which only goes back 50 years, coincides with the greatest peak in solar caused global warming, and coincides in addition with overlapping CO_2 emissions from historic high CO_2 concentrations via the slow moving global recycling system.

89

Nor was it ever said that CO2 isn't a climate factor anyway, as it contributes only a millionth part to the global greenhouse effect, which itself isn't the prime factor either, since the absolute prime factor is the rate of cloudiness that is affected by solar activity, which in turn is affected by astrophysical factors that flow through the Primer Fields.

Since all these factors are known, the political will is falsely placed and evidently with intention, especially considering that it is known that CO2, which the manmade global warming dogma is centered on, is not a climate factor at all, never was, and never can be as the claimed effect is physically impossible in the overall context.

The Great Global Warming that pulled us out of the Little Ice Age

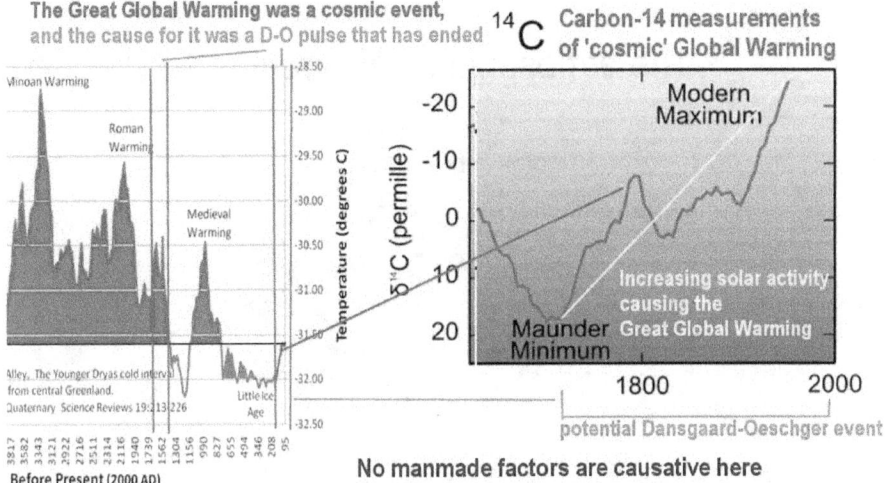

The Great Global Warming was a cosmic event, and the cause for it was a D-O pulse that has ended

Minoan Warming

Roman Warming

Medieval Warming

Alley, The Younger Dryas cold interval from central Greenland. Quaternary Science Reviews 19:213,226

Little Ice Age

Before Present (2000 AD)

Temperature (degrees C)

^{14}C Carbon-14 measurements of 'cosmic' Global Warming

Modern Maximum

Increasing solar activity causing the Great Global Warming

Maunder Minimum

1800   2000

potential Dansgaard-Oeschger event

No manmade factors are causative here

Nor was it ever said in the political brainwashing projects that include to a large degree also the media, that the Great Global Warming that pulled us out of the Little Ice Age was demonstrably and measurably a solar-forced cosmic phenomenon, instead of being manmade in any way.

** the exploration continues

The CO2 subject is far bigger than a mere academic concern

The atmospheric Jet Streams

Polar Jet

Subtropical Jet

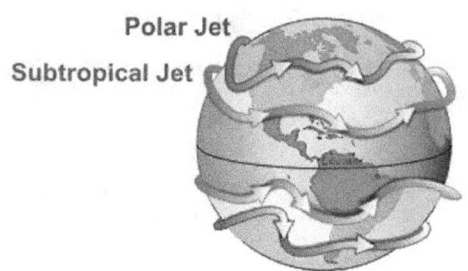

I have diverged into the CO2 subject, because it is a part of the global recycling system, which is powered by astrophysical principles that cause air and sea movements on Earth that flow faster than the rotation of the Earth. The CO2 subject, of course, is far bigger than a mere academic concern. It affects your dinner table with increasingly high food prices, and raises the fuel prices for transportation and for heating your home.

The mandated mass-burning of food for biofuels production

Under the banner of "CO2-forced Manmade Global Warming", the mandated mass-burning of food for biofuels production, has diverted huge agricultural resources to be burned, that would have normally nourished 400 million people. In a world that has a billion people living in chronic starvation, the food burning holocaust is claiming more than 100 million victims each year, of death by starvation.

The CO2 issue far out of mere academic concerns

PARIS2015
UN CLIMATE CHANGE CONFERENCE
COP21·CMP11

COP 21: Heads of delegations by GUSTAVO-CAMACHO-GONZALEZ - Licensed under CC BY 2.0 via Commons
by Presidencia de la República Mexicana -delegates

Poster of the Climate Conference.
Licensed under Fair use via Wikipedia

This takes the CO2 issue far out of the realm of mere academic concerns, into the political arena with the often stated objective of mass-depopulation, that is to weed out the 'useless eaters,' as the victims were once called, to shrink world population to less than a billion people. The academic issues, if they were widely understood, would block the politically-forced tragedies for which no physical cause actually exists.

*Getting back to the phenomena of fluid movements that are faster

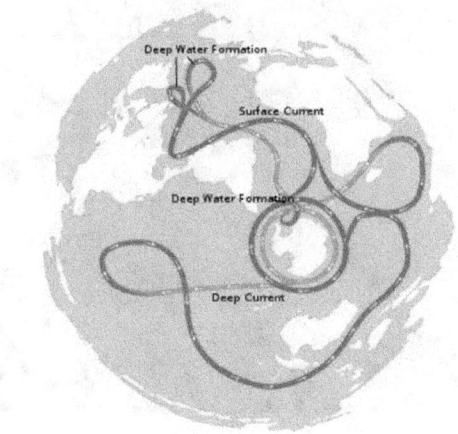

The ocean currents conveyor belt centered on the deep cold waters encircling Antarctica

"Conveyor belt" by Avsa - under CC BY-SA 3.0 via Commons -

In getting back to the phenomena of fluid movements that are faster and than the rotational speed of an object under the Primer Fields, as we see it reflected in the Jet Streams and the ocean-current movements around Antarctica, one finds of course, that the Earth is only a small part of the solar system, which means that it is not alone in being affected by the Primer Fields that cause fluid movements exceeding the spin-axis' rotational speed.

The Sun rotates significantly faster at the equator

The Sun in visible light
as seen through a dark glass

The phenomenon is highly prominent on the Sun, though it is barely visible there. The phenomenon is discernable by the differential movements of its feature, such as sunspots. The Sun rotates significantly faster at the equator than at the poles. It takes the Sun 35 days for a single rotation at the poles, but only 25 days for a rotation at the equator.

The faster equatorial rotation stands as evidence that the Sun is actively rotated by an external electric force, which in addition, acts most strongly on its equatorial region from where its heliospheric current sheet extends.

The Sun's two high-activity bands spaced centered off the equator

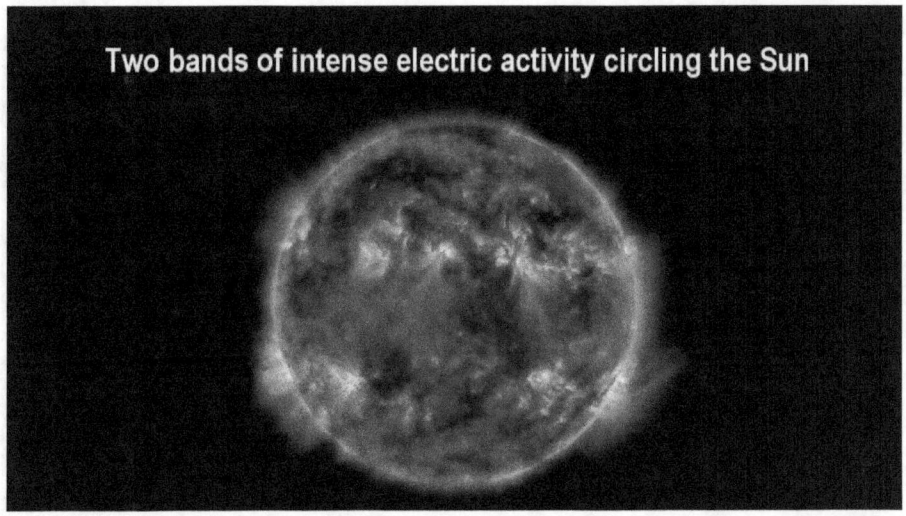

Two bands of intense electric activity circling the Sun

The Sun's two high-activity bands spaced centered off the equator, actually reflect in the large what became evident in principle in plasma experiments.

The evidence illustrates the functioning of the Primer Fields

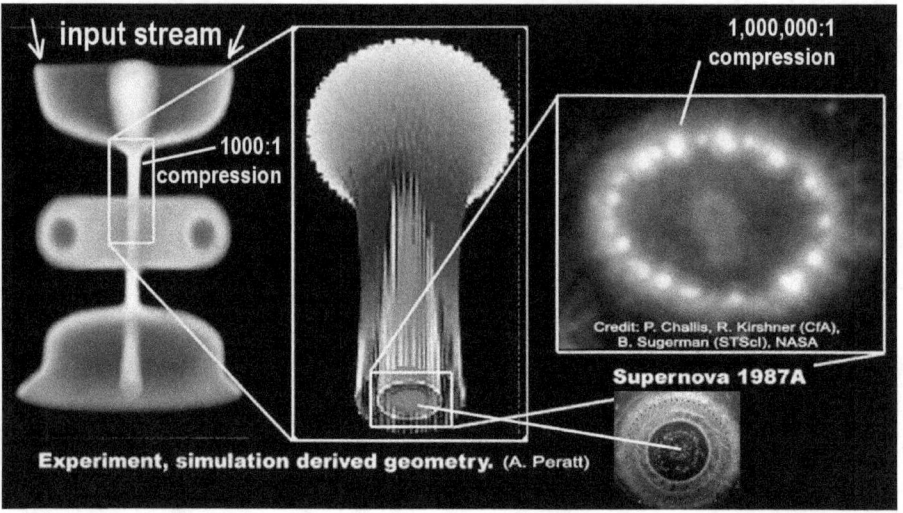

The evidence illustrates the functioning of the Primer Fields, and their effects that we see evident in the form of an ecliptic ring of induced plasma currents, which we see in principle reflected on the Sun in the form of the equatorial high-activity zones.

The ecliptic principle is also apparent

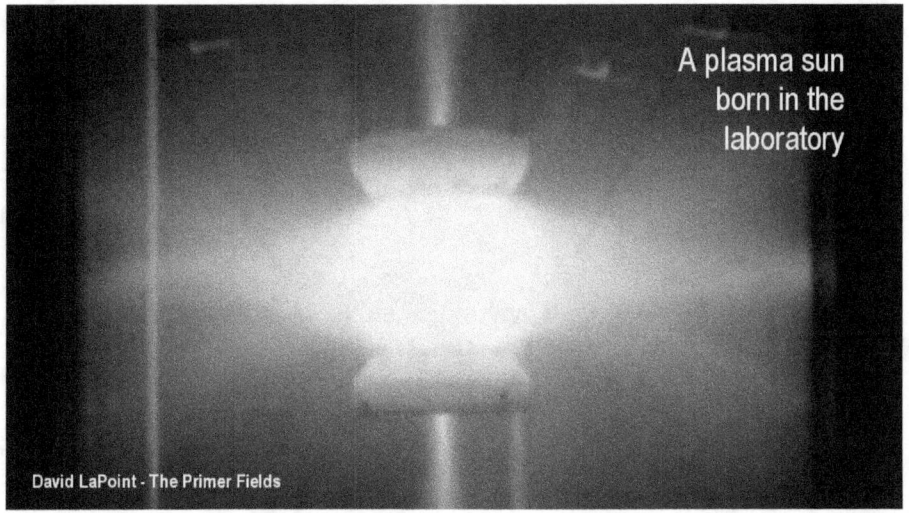

A plasma sun born in the laboratory

David LaPoint - The Primer Fields

The ecliptic principle is also apparent to some degree in David LaPoint's static exploration of the primer fields.

The resulting shape apparent in the ecliptic shape of galaxies

David LaPoint - The Primer Fields

He points out that the resulting shape in his experiments is dramatically apparent in the ecliptic shape of galaxies, which suggests a commonality of principle in both cases.

A solar system, that is in all major aspects electrically powered

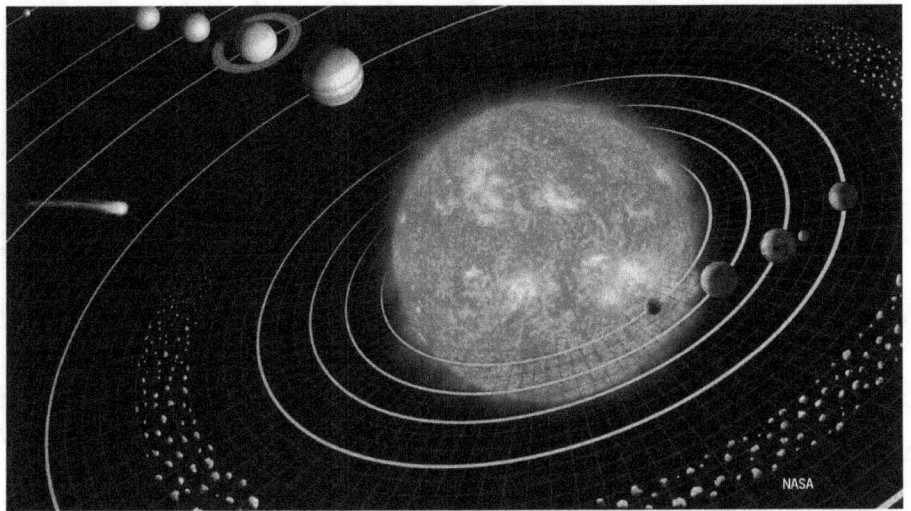

The evidence that one sees here is of a solar system, that is in all major aspects electrically powered and electrically motivated, and shares this principle with the galactic systems.

Galaxies at node points

This means that galaxies are not stable entities, but are subject to plasma-density resonances in intergalactic space, and intergalactic plasma streams that are focused by large primer fields onto the galaxies.

For our galaxy, the Milky Way Galaxy

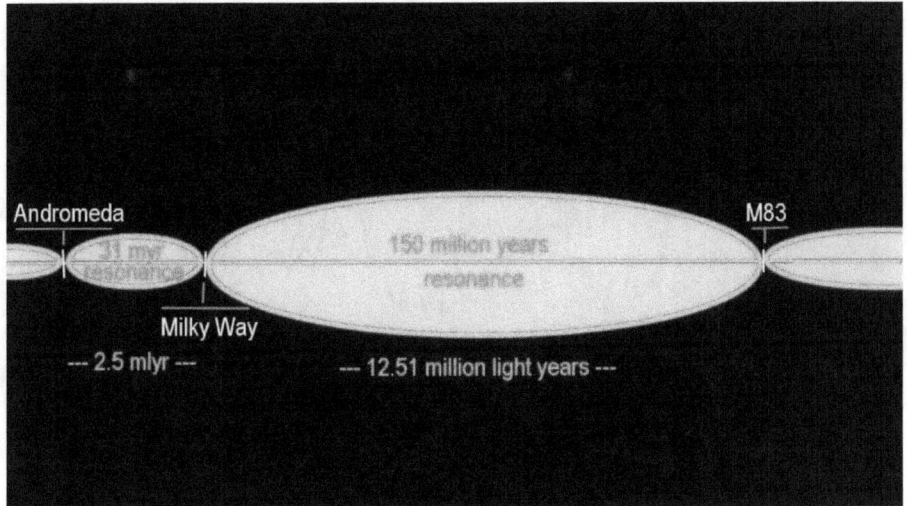

For our galaxy, the Milky Way Galaxy, two potential plasma streams to nearby galaxies come into view that form intergalactic node points. The potential resonance cycles of these very long intergalactic plasma streams, appear to be reflected in long-term climate cycles on Earth.

The two long intergalactic resonance cycles

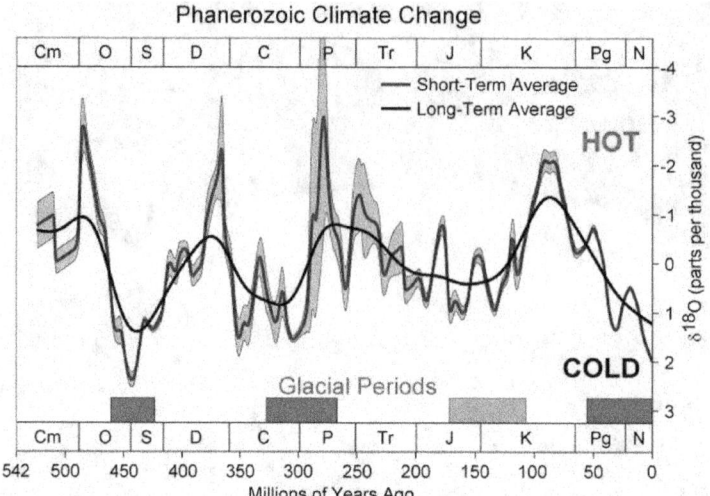

The two long intergalactic resonance cycles appear to reflect in combination the two overlapping climate cycles on Earth that we found evidence of in sediments, in the form of oxygen-18 isotope variations.

The solar system and the Earth are not fundamentally isolated entities

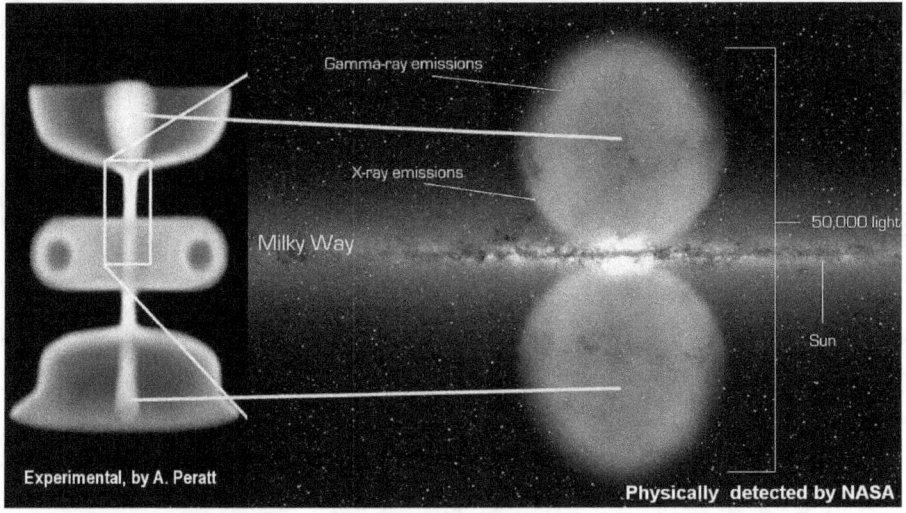

This means that the solar system and the Earth are not fundamentally isolated entities, but are intimately connected, both physically, and by the sharing of a common, universal, all-motivating principle.

This also means that the Sun is not its own master

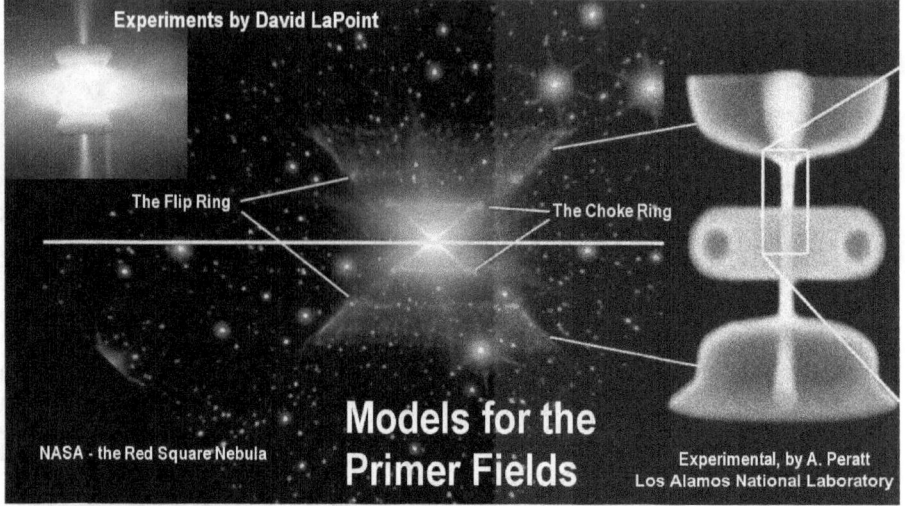

This also means that the Sun is not its own master, but exists cradled between interstellar primer fields that give it is existence as an externally powered star. This further means that the primer fields can collapse when the external conditions fail that enable the primer fields to function. When this happens, our world will change.

The phase shift in our world, onto an inactive Sun

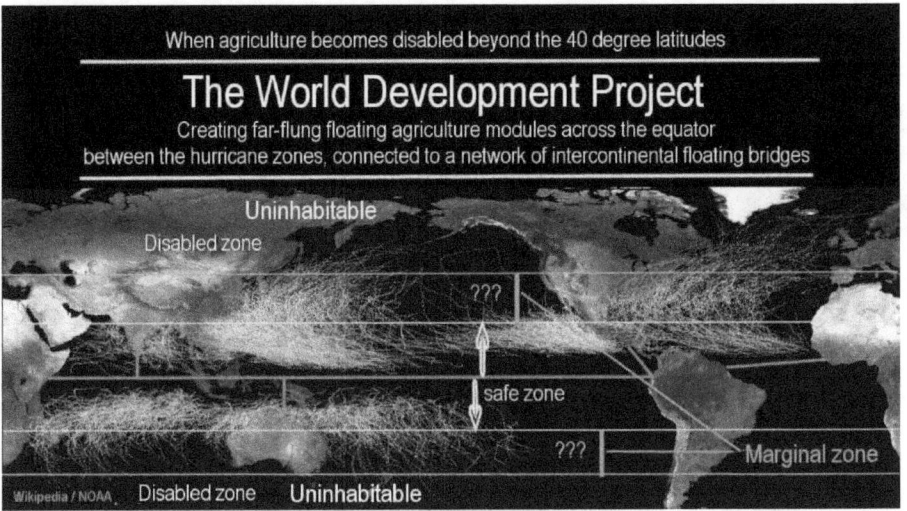

The phase shift in our world, onto an inactive Sun, which could happen in 30 years, does not imply that the end is near for mankind. It means the opposite. It means that we find ourselves impelled to utilize the scientific, technologic, economic, and cultural resources that we have developed, and meet the impending phase shift in the solar system with an equally majestic phase shift of our own, by creating the greatest renaissance of all times, in our time, against which the coming Ice Age has no sting for us.

www.ingramcontent.com/pod-product-compliance
Lightning Source LLC
Chambersburg PA
CBHW060356190526
45169CB00002B/619